NASTY, BRUTISH AND SHORT

Other *Quirks & Quarks* Books

The Quirks & Quarks Question Book
The Quirks & Quarks Guide to Space

NASTY, BRUTISH AND SHORT

The Quirks & Quarks Guide to Animal Sex and Other Weird Behaviour

PAT SENSON

Introduced by
Bob McDonald

McCLELLAND & STEWART

Copyright © 2010 by the Canadian Broadcasting Corporation

All rights reserved. The use of any part of this publication reproduced, transmitted in any form or by any means, electronic, mechanical, photocopying, recording, or otherwise, or stored in a retrieval system, without the prior written consent of the publisher – or, in case of photocopying or other reprographic copying, a licence from the Canadian Copyright Licensing Agency – is an infringement of the copyright law.

Library and Archives Canada Cataloguing in Publication

Senson, Pat, 1969-
Nasty, brutish, and short : the Quirks and quarks guide to animal sex and other weird behaviour / Pat Senson.

ISBN 978-0-7710-7968-9

1. Reproduction. 2. Animal behavior. 3. Animals. I. Title.

QP251.S45 2010   573.6   C2009-905210-5

We acknowledge the financial support of the Government of Canada through the Book Publishing Industry Development Program and that of the Government of Ontario through the Ontario Media Development Corporation's Ontario Book Initiative. We further acknowledge the support of the Canada Council for the Arts and the Ontario Arts Council for our publishing program.

Published simultaneously in the United States by McClelland & Stewart Ltd., P.O. Box 1030, Plattsburgh, New York 12901

Library of Congress Control Number: 2009935662

Typeset in Palatino by M&S, Toronto
Printed and bound in Canada

This book is printed on acid-free paper that is 100% recycled, ancient-forest friendly (100% post-consumer waste).

McClelland & Stewart Ltd.
75 Sherbourne Street
Toronto, Ontario
M5A 2P9
www.mcclelland.com

1 2 3 4 5   14 13 12 11 10

CONTENTS

INTRODUCTION — xi

1. THE BATTLE OF THE SEXES — 1

THE INSECTS
Insect Arms Race — 2
Weevil Penises — 5
The Power of One — 7
Bedbug Penis Tasting — 10
Fishing Spider Sex — 12
Redback Spider Sex — 15
Some Crickets Like it Rough — 19

THE OTHERS — 21
Genitalia Most Fowl — 21
Cannibal Fish — 24
Infertile Monkeys — 26
Meerkat Infanticide — 29
Hyena Hormones — 32
Speedy Sperm — 34

2. IMPRESSING THE OTHER SEX — 37

BIRDS DO IT
Zebra Finch IQ — 38
Bowerbird Bullies — 41
Bowerbirds Come on Strong — 43
Godwit Vacations — 45
Auklets and Aphrodisiacs — 48
The Sound of Two Wings Clapping — 50

BEES (AND OTHER INSECTS) DO IT

Hot Insect Sex	53
Quiet Crickets	56
Sniffing for Sexual Satisfaction	58
Weta Walking	60
Cricket Coupling	63
Brotherly Love Among Crayfish	66
Bisexual Beetles	69
Cockroach Wars	72
Water from Weevils' Sperm	74
Lying Flies	76

EVEN ANIMALS, LAND AND SEA, DO IT

Primate Copulation Call	79
Salacious Simians	81
Bat Brains and Balls	83
Amorous Antelope Antics	86
Sea Slug Orgy	88
The Barnacle's Penis	91
Penis Preference in Mosquitofish	94
Amazon Mollies	96
Dolphin Bling	99
Cichlid Dominance	101

3. Parenting Skills — 105

RAISING THE CHICKS

Cowbirds' Bad Influence	106
There Goes the Neighbourhood	109
Birds that Time-share	112
Two-Mom Albatross Families	115

BRINGING UP A BROOD

Self-Roasting Rodents	118

Father's Nose Best	121
Pregnancy and Monkey Dads	123
Stinking Baby Bugs	126
Cannibal Tadpoles	128

4. LUNCH, ANYONE? 131
FINDING FOOD
Owl Dung Nests	132
Snails and their Slime	135
How a Bat Holds its Licker	137
Its Bite is Worse than its Bite	140
Whales Lunge for Lunch	141
Killer Whale Sonar	144
Sea Lion Diving	146
Dolphin Herders	150
Guano-Eating Salamanders	152
The Hummingbird's Stopwatch	154

HOW TO AVOID BECOMING A MEAL
Wolverine Frogs	157
Bombardier Beetles	159
Toadfish Pee	161
Squirrels Eating Snake Skin	164
Ultrasonic Gophers	167
Crafty Chameleon Camouflage	169

5. PARSING PARASITE PECULIARITIES 173
Blister Beetle Deception	174
Ants Look Berry Nice	176
Crustacean Kids, Parasitic Parents	179
Woolly Caterpillar Medicine	181

6. Higher, Faster, Stronger — 185

- The Fastest Jaw in the South — 186
- Preying Mantis Shrimp — 189
- The Mantis Shrimp's Super-Sight — 192
- Mandible Mayhem — 195
- How the Bow Bug Shoots Itself — 196
- Catapulting Caterpillar Feces — 199

7. Even More Weird Behaviour — 203

- Washing with Urine — 204
- Spider Venom Side Effect — 207
- Termite Head-Bangers — 209
- Army Ants Lie Down on the Job — 211
- Loony Tunes — 214
- Electric Fish — 216
- Chimp Plans for Pitching — 219
- Crocodiles Go Home — 222
- Alligator Air Bags — 225
- Bird Doping — 227
- Scuba Bug — 229
- Head-Under-Heels — 232
- Bees' Knees — 235
- Clownfish Sex — 237
- Worm Hibernation — 240
- Four-Eyed Fish — 242
- Cold, Dry Bugs — 245
- Tonguefish at the Vents — 248

8. The Games Researchers Play — 251

- Spitting Cobras — 252
- Worm Grunting — 255
- Flatulent Fish — 258
- Fearless Iguanas — 260
- Running on Water — 262
- Wasp Faces: Friend or Foe — 265
- Fruit Fly Fight Club — 267
- Tadpole Bail Out — 269
- Flying Ants — 272

Introduction

Let me see if I've got this right: the male spider rips off part of his huge sexual organ, so he can run faster, in order to catch a female who is many times his size. A bean weevil inserts a barbed penis into his mate's reproductive tract, which tears her abdomen. She, of course, fights him off, but needs his sperm. Another type of female spider tries to bite the head off the male, while he is mounting her, which only increases his vigour. And if he isn't quick enough, he becomes her lunch when the mating is over.

Aren't you glad you're not an insect?

Maybe it would be better to just clone yourself without a partner. Well, that's okay, too. In fact, anything is possible in the weird and wonderful world of animal sexual behaviour.

Over the years on *Quirks & Quarks* (the CBC Radio program that I host), scientists have told us about the extraordinary range of bizarre adaptations nature has come up with, to make sure that birds, bees, and every other form of life can "do it" successfully. And although we're all both fascinated and repulsed by stories of strange animal mating, it's not just the sexual behaviour that captures scientists' attention and scrutiny. They've also told us stories about the formidable weapons, stealthy strategies, clever tricks, and just plain weird means that animals use to find food or avoid being eaten. Each time we do one of these interviews, we come out of the studio smiling and shaking our heads in wonder. So, we decided to gather together almost a hundred of these tales from over the last nine years, and put them together into this book for you.

In the animal and insect world, the whole idea of sex is to make sure your genes, and not someone else's, get passed on to the next generation. That means fighting off competition and finding the best mate. Of course, this results in lots of head-butting among males and deception by females, and sometimes you just have to wonder how some animals get it on at all.

Take the case of the Argentine lake duck. Not only is the male well hung, his organ is shaped like a corkscrew. And if that's not strange enough, the female reproductive tract is also spiral-shaped, but in the opposite direction. That's like trying to screw a bolt with the wrong thread into a nut. How do the ducks do it? Read the first chapter and find out.

Speaking of well hung, the world record-holder for the longest penis goes to the humble barnacle, with an organ four times longer than its body. Fortunately, the animal doesn't have to worry about walking around with something that long dragging along the ground, because it can't walk. The organ goes out looking for females on its own. You can discover the delicious details in the second chapter of this book.

Of course, it's impossible to hear stories about animal sexual behaviour without thinking about how they reflect on our own human activities. For instance, the bower bird builds a "bachelor pad" to impress the girls and even does a dance in front of it, to entice them in. Dolphins dress themselves up in "bling" made of leaves and sticks during mating season; and the godwit birds in the Arctic mate for life – but take separate vacations (to keep the relationship strong?). They're all in the book as well.

And don't think the Animal Kingdom sticks to traditional relationships, where huge, dominant males rule over a harem of subordinate females. Among hyenas, for example, females are the biggest in the pack and even carry a penis-like organ that they use to keep the males in line. Beetles can be bisexual, and hermaphroditic sea slugs have group orgies where they show up as males and change to females before the party's over. And we thought the Hippies invented free love.

Besides outperforming us on the sexual stage, animals also outperform even our most talented athletes, with amazing feats of strength and speed. A caterpillar throws its feces the equivalent of a seventy-six-yard field goal (you'll have to read Chapter 6 to find out why). The mantis shrimp throws a ninety-kilometre (fifty-six-mile) per-hour punch – underwater. And the froghopper bug shoots itself through the air, using a built-in bow-and-arrow mechanism that accelerates it upwards at 400 Gs. That's more than a hundred times faster than the Space Shuttle. The Olympics are tame, by comparison.

The scientists who study these creatures are constantly amazed by what they see through their high-speed cameras and microscopes. A bat flicks out a tongue with bristles on the end to catch an insect. It's too fast to see. But in slow-motion replay, they discover the tongue is so long that it would be the equivalent of a human tongue sticking out two metres (two yards) in front of your face. Then, miraculously, the tentacle-like organ folds down between the bat's rib cage and the heart, ready for the next strike.

Other scientists perform some strange and often comical antics to get their results, such as inflating a dissected barnacle penis with a syringe, to see how long it is, or taunting a cobra from behind Plexiglas, to see how it shoots its venom. (The cobra and the intrepid researcher star in Chapter 8.)

Science fiction writers have been outdone by the creativity of the animals. A tropical frog has sharp claws that magically pierce through its skin, just like the Adamantium blades of Wolverine, a character in the *X-Men* series. The moray eel has a second set of retractable jaws in its throat, similar to the killer jaws of the creature in *Alien*. And those are just two examples of formidable weaponry wielded by tiny creatures.

Even engineers would be impressed by animal innovations. The bombardier beetle mixes volatile chemicals in its abdomen that instantly produce boiling water and steam, which the insect shoots out, like a rocket engine. (The German buzz bombs that rained down on London in World War II used the same propulsion principle.) Opticians will be familiar with a fish eye that uses a mirror instead of a lens to produce an image of the dark ocean floor, exactly the way the Hubble Space Telescope uses a mirror to capture images from the depths of space.

In 2009 we toasted the Year of Darwin (the 200th anniversary of his birth) by celebrating the wonders of natural selection and the amazing innovations that evolution has produced to enable life to fill every niche on the planet. Today's scientists continue that tradition by looking under leaves and rocks, crawling into bat-infested caves, and plodding through mosquito-infested swamps to get a glimpse into the fascinating non-human world that surrounds us. There are an estimated 50 million different species of animal on the planet. So far, scientists have identified only about 1.5 million of them.

But zoologists, entomologists, and wildlife biologists have their work cut out for them. Thanks to human activities, we are in the midst of an extinction event in which species are disappearing at a rate not seen since the last time a large asteroid hit the Earth – 65 million years ago. We're destroying their habitat, paving over their food sources, and hunting and fishing them at an alarming rate. These animals, with all their bizarre and wonderful behaviours, must be catalogued and understood before they are gone forever.

So, the next time you're in a forest or a meadow, lie down and get your face really close to the ground. Take a look between the blades of grass, under the leaves and rocks, or even dig down a bit into the soil. You will find a miniature zoo, populated by creatures as alien as those in any science fiction movie. Welcome to the rest of the world. And if you manage to see some of the inhabitants scurrying about, remember: they have lives, they have families, and they have amazing abilities. And we probably couldn't survive without them.

I've always been a fan of space exploration and the exciting possibility of alien life on other worlds. But we can already explore strange alien-looking creatures, with unusual lifestyles, by simply stopping to take a close look at the alternate universe that thrives beneath our feet, over our heads, and in every place on Earth you can imagine. Read on, and get ready to be amazed by life in all its glorious (and bizarre) splendour.

Thanks to former *Quirks & Quarks* producer Pat Senson for turning the scientists' often complex and difficult descriptions into a lively and comprehensible text. He and veteran producer Jim Lebans produced most of the interviews on which this book is based. Without them, we wouldn't have *Nasty, Brutish and Short*, or *Quirks & Quarks*, for that matter. And thanks also to senior producer Jim Handman, who conceived the book, and has always said that what the program needs is more sex with animals. Enjoy.

Bob McDonald
September 2009
Toronto

EDITORIAL NOTE

All quotes from the scientists in this book are taken from radio interviews they gave to *Quirks & Quarks*, the national science program of the Canadian Broadcasting Corporation. The interviews were conducted by Bob McDonald, the host of the program, between 2000 and 2009. We have tried to give the current academic affiliation for each scientist, whenever possible, as of summer 2009. In some cases, they were affiliated with other academic institutions at the time of the interviews.

1

THE BATTLE OF THE SEXES

The Insects

❖

For examples of extreme sexual conflict in the animal world, insects are the place to look. From arms races to fatal copulation, they do it all.

Insect Arms Race

It would hardly be novel to say that males and females frequently want different things. But when it comes to insects, we're not talking about hanging out with buddies at a football game versus staying home watching romantic movies. No, this is a story about sex, and so it's ultimately about evolution, with males wanting one evolutionary outcome and females another. Nature has set up some species for sexual conflict that goes far beyond the battle over who takes out the garbage. Take, for example, the water strider. This group of insects is locked in a sexual struggle that has all the features of an evolutionary arms race, complete with occasional detente and the threat of mutually assured destruction.

You would probably recognize water striders if you saw them. They're found all over the world, and according to Dr. Locke Rowe, an evolutionary biologist at the University of Toronto, just about anywhere there's a pool of water there will be water striders. As their name suggests, they're able to walk across the surface of the water, standing on the ends of their long legs.

Their sexual conflict arises out of the different desires of the males and females. The males want as many matings as possible. The more they mate, the more babies are their kin, and the more genes they'll pass on. For the females, though, mating is much more costly, since it takes a lot of energy to produce eggs. As soon as a female's eggs are fertilized, she doesn't need to mate again, but the males are still going to show interest. Also, as Dr. Rowe says, "The females actually carry the males during mating, and that's an energetic cost. So they fight. If you see water striders on the surface and look closely, you'll see, after some time, males jumping on females and females struggling and somersaulting to get rid of them."

So, the females have come up with ways to try and keep the males off their backs once they have already mated. In some species they have developed long spines that make it difficult for the males to climb on them. The males have fought back by evolving their entire body into one big grappling device. Each of their three pairs of legs has grappling hooks and spines that allow them to hang on, for dear life, to the females. In one species, the males have evolved antennae that are, Dr. Rowe says, "big, muscularized, swollen-up armaments that they use for grasping those females." These antennae are not much use for sensing the environment any more (their original purpose) – but very helpful when the female doesn't want to co-operate.

This back-and-forth retaliation has been going on throughout the water striders' evolution. The females develop a way of repelling males, and the males come up with a way of countering the new apparatus. Then the female comes up with a new defence, and the male responds. And so on. Except, not always. Dr. Rowe has observed that sometimes there's a de-escalation in the arms race, and there are fewer and fewer armaments on successive generations of the insects. Depending on the particular species, the arms race seems to be raging, quieting down, or staying steady. Curiously, as long as there's a balance between the two sexes' armaments, all these different species are about equally successful at reproducing. That's a lot like a human arms race. As long as everyone's in the same boat, then how well armed you are isn't the point; it's the balance that counts.

Sometimes the balance is off, and whenever the male has the advantage, mating rates shoot up to as much as ten times what they are in other strider species. But that puts females under a great deal of pressure to evolve defences, so it isn't long before they do and balance is restored. This explains how escalation can happen, but why there's de-escalation is more complicated.

No one knows for sure, but Dr. Rowe thinks it's because of the heavy costs of the arms race to the individual insects. Their armaments take a lot of energy to grow and maintain and can get in the way of normal activity. So, if they are not needed, might as well get rid of them. De-escalation happens when females are so far behind the males in the arms race that they just give up. Then the males don't need to be so heavily armoured, and the ones who waste less energy building grappling hooks have the advantage, and so gradually both sexes scale back. Same thing when males are scarce; then the females' desire to breed leads them over a few generations to abandon their weapons and be nice to the guys.

But whether it's a full-scale war or just a skirmish, as long as the armaments on both sides are balanced, then this arms race helps the species keep striding along.

WEEVIL PENISES

It's pretty safe to say that men and women generally have different priorities when it comes to sex. But that isn't just a human trait. Throughout the animal kingdom, what females want is not always matched by male desire. A good example is the humble bean weevil (*Callosobruchus maculatus*), an animal that seems to have taken the battle of the sexes to a new extreme. Their copulatory conflict was first described by Dr. Helen Crudgington, a researcher at the University of Sheffield in England.

Looking for sexual activity among bean weevils is no simple task. The beetles are a common pest species found throughout the world, but they're really small, only about three and a half millimetres long (about an eighth of an inch). And they're fairly nondescript brown bugs, or they appear to be until you look at the males really closely. Examine the penis of a bean weevil and you'll discover it's covered in hard, sharp spines. The purpose of these spines? Well, when a male bean weevil mates, these spines puncture the lining of the female's genital tract.

This is not, to say the least, what a female is looking for in a sexual encounter. So, she's faced with a problem. She does need to get fertilized. After all, as Dr. Crudgington explains it, "It may not be in the interests of females to go along with those matings, but they obviously need at least one to get sperm to fertilize their eggs. And in evolutionary terms, it's no good if you remain a virgin." But the damage caused by the males, says Dr. Crudgington, "can produce costs for the female in terms of dehydration – that is, losing moisture from these wounds. And they can also be costly in the sense that they can be a route for the entrance of harmful pathogenic organisms. So, obviously, this is not ideal for the females." Now there's an understatement.

From the male perspective, however, there are two possible advantages to damaging the female like this. First of all, if she is injured, she's unlikely to try to mate again until she's healed. In the meantime, her eggs will mature, and the male who has fertilized her gets to be the father of all the offspring. Second, these wounds are so traumatic that they can leave the female close to death, and the biological response to that is to produce a lot of eggs quickly. It could be her last chance to have offspring, so, from an evolutionary perspective, she wants to have as many as possible, as soon as possible. And the male who got to her first is the one whose sperm she's going to use when she lays those eggs. But it's not in the male's best interests to do too much damage to the female. Killing her is no good – she won't produce any offspring that way, and he'll have wasted all that energy chasing her down. For the male bean weevil, there has to be some level of restraint.

So we end up with a system that requires a certain balance. The male wants to mate with, but not damage too much, as many females as possible. The females want to mate with as few males as they can. How this manifests is quite curious. It's what got Dr. Crudgington to look at the bean weevil's sexual practices in the first place. She had noticed that, during mating, the females furiously kick at the males to get them off their backs. This "mate kicking," as Dr. Crudgington calls it, is quite awkward, since the spines on the penis are lodged inside the females. But it seems to work. The females minimize their injuries, the males unload their sperm, and the species is thriving.

There's an evolutionary lesson in all this, though. It doesn't seem to make evolutionary sense for a species if sex is this dangerous for the individuals involved. It can only reduce the number of potential offspring. But evolution doesn't work at the level of species. It's the pressure on the individual to do the best he or she can for himself or herself that results in evolutionary change. And these spiny penises are great for the males, so they've stuck around (so to speak).

THE POWER OF ONE

Some stories are not for the squeamish, and this one might leave you on edge. So, take a deep breath before you read on.

There's a spider, *Tidarren sisyphoides*, that has an unusual sexual habit. While getting ready to find a female and mate, the male rips off one of his two sex organs and casts it aside. Then, unencumbered by the weight of the organ, he heads off in pursuit of a paramour. This seems like an odd and extremely painful measure, but Dr. Duncan Irschick, now a behavioural ecologist at the University of Massachusetts, Amherst, thinks he knows why self-mutilation is this spider's best strategy.

The answer comes from the evolutionary history of this critter. As is the case with many species of spider, the female *T. sisyphoides* is much larger than the male. In fact, in this species, the difference is huge: the female is one hundred times larger. And, not to put too fine a point on it, her sex organs are proportional to her body size. That's a real problem for the male. If he's one hundred times smaller than his mate, and his sex organs – or pedipalps, as they're called – are proportional to his body, he's never going to match up with his partner. What's he to do? The answer is to develop really big pedipalps.

Calling them "really big" is an understatement. The two pedipalps each make up about 10 per cent of the spider's total body mass. Dr. Irschick describes them as "giant beach balls sitting in front of the male." They are big enough to fertilize the female, but they come at a cost for the males. They're so large that when the male starts to walk or run, they drag along.

The male spiders want to find the females as quickly as possible. After all, as Dr. Irschick says, "Whichever male in this species finds the female first gets all the goodies." If you want to speed yourself up, as any car enthusiast knows, one way is to reduce drag. So, these male spiders spin a silk thread (they are spiders, after all), wrap it around one of their pedipalps, tighten the thread to seal it off, suck out the fluid, and, well, rip off the unnecessary organ. Don't try this at home! Also, as Dr. Irschick discovered, humans shouldn't do this to the spider – he won't survive. But when the male performs the surgery on himself, he's able to head off quickly in search of a mate.

Removing this organ makes a major difference. In Dr. Irschick's lab (he was working at Tulane University at the time), the researchers chased spiders, some with one and some with two pedipalps, around a track, to see how long they'd last. On average, the "intact" males pooped out after sixteen minutes, while the single-sided spiders lasted for as long as twenty-eight. Putting it in human terms, Dr. Irschick says, "That's the equivalent of you going out and running about 2 miles [3.2 kilometres] versus ripping off your reproductive organ, or leg or some appendage, and then going out and running 3.5 miles [5.6 kilometres]. It's a big performance advantage."

The males definitely need this advantage. Female spiders of this species are few and far between in the forests of the southern United States, where they live. So males have a real challenge finding them. Any benefit in speed that they can get for their hunt is going to be critical. Of course, this advantage might come at a cost, too. It's possible that losing one pedipalp cuts down on the amount of sperm the male can produce. But as far as anyone can tell, the males seem to be doing all right so far – the species' population isn't in any trouble. And while it might be painful for the spiders, until someone comes up with a way to conduct spider questionnaires, we'll never know for sure whether it is.

If you think that the male has already sacrificed enough for the sake of sex, think again. For the male of this spider species, mating is basically a suicide mission. After he finds a female, they copulate and he stays attached to her body, which probably blocks access for other males. Eventually he dies, at which point, Dr. Irschick says, "He basically gets sucked dry [by the female] and flicked off like a piece of dirt, afterwards." This is not exactly a great future for a young male spider to look forward to, but at least his genes are transferred to the next generation.

There's one question that remains in this tale of self-sacrifice. Wouldn't it be much easier for the male to evolve a larger body, to mate more easily with the female and keep his organs from dragging? The problem is, the bigger the individual male, the more energy he needs to expend finding a female. So, in fact, being small is an evolutionary advantage to these males. For the females, however, bigger is better, since the larger she is, the more eggs she can produce. So, the organ removal by the males seems to be a compromise solution. The alternative of evolving a single pedipalp is probably too difficult, because biology prefers symmetry. And ripping one off is a quick, easy operation, and it gets the job done.

But it does make you wonder: what possessed the first male spider to try tearing off a part of himself?

Bedbug Penis Tasting

Bedbugs are not a particularly romantic lot. For the males, sex is *wham, bam, thank you ma'am*. The females, on the other hand, are a deceptive bunch, each trying to appear chaste, even virginal, in their attempt to attract the male with the most sperm. But male bedbugs have their own means of testing a lady's virtue. They've developed a phallus that can detect whether a female they've mated with has been faithful. Dr. Michael Siva-Jothy, an evolutionary physiologist at the University of Sheffield in England, thinks that this is an adaptation that prevents the males from wasting sperm.

Before we get to the perceptive penis, here are some other interesting facts about bedbug sex that are worth knowing. First, even though the females have a fully developed reproductive tract, the males don't use it. It's strictly for laying eggs. Dr. Siva-Jothy explains this oddity: "The male's penis is like a hypodermic needle. He stabs it into the body wall of the female, and he inseminates directly into her body cavity. Now, that's pretty strange." Pretty strange, indeed. The sperm migrate into the female's bloodstream and swim to the ovaries, where they fertilize the eggs.

This isn't some kind of evolutionary mistake. Nor do the males just have really bad aim. In fact, while the male bedbugs of primitive species will stab females anywhere in the abdomen, Dr. Siva-Jothy says that in the more evolved species "the females have little grooves and channels in their cuticular armour, which guide the male to one particular spot." This spot covers an organ, unique to bedbugs, that stores immune cells which protect the female from infection. This is a pretty important defence for her, since she has to put up with being stabbed by any interested males.

This is where the sensing penis comes into the picture.

Dr. Siva-Jothy noticed that when females mate with more than one male (not an uncommon occurrence in the bedbug world), the first male copulates for longer, and deposits more sperm than the males who come along later. In fact, the last male to stab her would be the fastest at it and leave the least amount of sperm. Since the only part of the male bedbug that comes into direct contact with the female is the penis, it set Dr. Siva-Jothy to wondering if there was something special about this organ that was allowing the male to tell whether the female was a first-timer or not.

Interestingly, a look at a bedbug's penis under an electron microscope reveals that it's covered in tiny, fine bumps, which in other insects are associated with chemical receptors – or, as we think of them, taste buds. It turns out that a bedbug's penis is capable of tasting the female's abdomen.

To test whether this was indeed what was going on, Dr. Siva-Jothy painted the penises of some virgin male bedbugs with sperm from other males and let them breed with virgin females. (And if you want to know how to paint the penis of a bedbug, Dr. Siva-Jothy's says, "With a paintbrush.") Sure enough, the males with the sperm-painted penises spent less time mating, and deposited less sperm, than the males whose penis had been painted with salt water. They were "tasting" or sensing the alien sperm on their penis and surmising that another male had got to the female first, even though these females still had their virtue intact.

Why would they do this? While the answer isn't completely clear, it most likely has to do with sperm competition. All the males who mate with one female want their sperm to fertilize her eggs. But it appears that when the first male mounts, most of his sperm are absorbed by her defensive cells. Males who arrive later don't need to deposit as many sperm, as the cells are all full of the first male's sperm and his can get past and carry on their journey.

In fact, the last male has about a 68 per cent advantage over his earlier rivals. His sperm will fertilize 68 per cent of the eggs, even though he supplies the least amount. And he's got sperm left over for another mating with another female.

The advantage to the females of this weird way of mating isn't at all clear. But if they're going to get attacked during sex, at least they've found some small way to protect themselves. It's sexual competition at its most fierce!

Fishing Spider Sex

The *femme fatale* was a staple of classic Hollywood movies even before Barbara Stanwyck used her charms on a hapless insurance salesman to shocking effect in *Double Indemnity*. And the men in those movies never learn. The allure of sexual danger is just too much for them to resist. But even Hollywood's greatest man-eaters have nothing on nature's most dangerous vamp, the female fishing spider (*Dolomedes triton*). These females are so ferocious that when they invite a male for dinner, he often ends up as the appetizer. That's what Dr. Chad Johnson, a researcher at Arizona State University, has found, and he thinks he knows why males sometimes end up as dinner, and sometimes as a date.

Fishing spiders are found in lakes and ponds throughout eastern North America, as well as in Central and South America. Like the water striders that most people have seen, these critters can run on the surface of water, but they can also dive underneath it. Yet they spend most of their time at the water's edge, anchored to a plant, where they wait for vibrations on the surface. When they feel a ripple through special sensors on their legs, they pounce, capturing whatever hapless insect – be it grasshopper, cricket, or damsel fly (or leaf or twig) – has alighted nearby.

This is fascinating behaviour, but, of course, what really interests us is how males and females copulate. In many spider species, the male takes a risk when he mates. After the deed is done, it's not unusual for him to be eaten by the female. But Dr. Johnson has noticed something different in this species. "Fishing spiders are known for their voracity," he says, "in that females will attack courting males before the male has even mounted the female. Females will take a lunge at him." Dr. Johnson describes this as a "conundrum" for the spiders. Obviously, males aren't getting anything out of the interaction if they're ending up as food before they've had a chance to mate. But then, the females aren't getting anything other than a meal either. They are not getting their eggs fertilized. And the species isn't going to survive for very long if they eat every male that comes along. The female has to have some means of selecting when to eat males and when not to. That's what Dr. Johnson wanted to figure out.

The first idea he tested was something called the "adaptive foraging hypothesis," he says. "That's a very intuitive idea that sometimes food might be more important to the female than sperm. If that's the case, then the females might benefit more by eating the male than by mating with him." Of course, this doesn't explain why the female wouldn't try to do both, mate then eat, but it could be that food is the more pressing need for the female at the time.

The hypothesis didn't hold up. "Hungry females," Dr. Johnson says, "do not try to attack males more often than well-fed females, and that would be an underlying assumption of this adaptive foraging hypothesis." Hungry or not, the females' desire to chomp on males didn't change. Which led Dr. Johnson to turn the hypothesis on its head.

The flip side of the idea about food availability is male availability. If females sense that there are plenty of males around, then sacrificing one or two for food might not be such a bad thing. Another one will come along soon enough, and when she's ready to mate, he'll be by. On the other hand, if there are few males out there, then it's better to mate first and eat later, just in case she never sees another male. Knowing how many males are around, then, becomes key.

There's some evidence that this might be happening for fishing spiders in the wild. Male fishing spiders mature before the females and will move into an area where a female lives and try cohabiting with her until she matures. Dr. Johnson says, "This is usually two or three weeks, in which he's living two or three cattails over. They can smell each other; they can hear each other's vibrations on the water. He doesn't get too close because she's cannibalistic at this stage as well, but he does give the female some experience of his presence. I've even found females in the field cohabiting with two or three males." All this suggests that young females might know whether they're in a good place for finding males.

Dr. Johnson tested this idea out in the lab (he was working at the time at the University of Toronto, Scarborough, with Dr. Maydianne Andrade, whom you'll meet in the next story). "I housed some females with cohabiting males," he says, "and I housed other females with controlled females – this is, another spider that won't give them the pheromonal, or chemical cue. So the prediction was that females housed with males would obtain an experience of future mate availability and they would actually be more likely to attack males upon adulthood than females that were housed with another female." And that's exactly what he found. The females housed with males were much more likely to eat males before sex when they were adults, showing that it's a judgment they make based on earlier experience. Lots of males around, then she might as well eat a few before settling down to mate.

Unfortunately for the males, they have very little control over this. The females are larger than they are, by up to one third. So the males, who want to breed, have to take this risk. Sometimes they get lucky, other times they just end up as lunch. In this species, familiarity, when young, doesn't breed contempt – it breeds an appetite.

Redback Spider Sex

Humans, for the most part, have transformed the sex act from a mere biological necessity into something playful and pleasurable that's now usually not about reproduction. We aren't the only creatures to do this, but for many other species, sex is a deadly serious activity. Literally deadly, in some cases. For instance, we know that many male spiders get eaten by the female after mating. It's what biologists call sexual cannibalism. And there's one species, known as the redback spider (*Latrodectus hasselti*), that's added a particularly brutal twist to their mating ritual. For them, sex and sustenance are tightly intertwined, as Dr. Maydianne Andrade, an evolutionary biologist at the University of Toronto, Scarborough, has discovered.

Redback spiders are native to Australia, and they look a lot like their North American cousin, the black widow. The main difference is the red stripe, missing on the black widow, but present on the back of the redback. Females of this species are one centimetre long (a third of an inch, or about the size of a marble); males are much smaller, at about three to four millimetres long (something like a seventh of an inch, or the size of a grain of rice). Don't step on one of these if you see them. Like the black widow, the redback is a poisonous character.

You've probably heard of the black widow, and even its name suggests that the female eats the male after sex. But in the redback spider, the female doesn't need to work nearly as hard to get her post-sex meal. "Instead," says Dr. Andrade, "males of this species actually twist their body onto the female's fangs while they're copulating. Because of their unique anatomy, the female can eat the male while they're copulating." That's right, the males not only allow themselves to get eaten by the females, they encourage it. Why? you ask. It turns out that being eaten alive gives a male a mating advantage.

For reasons that aren't yet clear, the mating lasts longer when the female is eating at the same time. The longer copulation lasts, the better it is for the male in two ways. First, the longer sex lasts, the more eggs he can fertilize; and second, if he mates long enough, then he can place a plug in the female's reproductive tract. That sounds harmless enough, except for one thing: the plug is a part of his own body. "That's where it gets even more gruesome," says Dr. Andrade. "Males of this species actually break off a piece of their copulatory organ and deposit it inside the female's genitalia." Why put in a plug? Quite simply, it prevents any other male from depositing his sperm in the female. Which is one way of ensuring your genes are the ones that get passed on to the next generation.

But the story doesn't end there. Female redbacks store the sperm in two separate sacs. This ups the ante for the male, as he wants to fill both sacs. This means he's got to go through the whole mating sequence twice. Luckily, even though he's already broken off one of his copulatory organs, known as palps, he's got a second one, and her reproductive tract is only partially blocked by his first palp. So, says Dr. Andrade, "He copulates with one and he's actually standing on the female while he's doing this. He then climbs off the female and courts her again, climbs back on and inserts his other copulatory organ."

Remarkably, the male spider goes at it the second time already partially eaten. The female was chewing away on the male during the first round, and continues to eat him for the second round. Because the male needs to survive long enough to get to the second copulation, he's evolved to make this possible, but it isn't pretty.

Once copulation starts, the male, whose abdomen looks like a swollen grain of rice when sex begins, contracts around the middle, as if tightening a belt. The female chews on one end, and it looks as if the male has forced all his organs to the other. She gets a meal, but he gets to survive. Dr. Andrade says, "When he does this, even though the female is biting into him and blood is bubbling out, he's able to court her again. In our experiments, we found that he could survive long enough to run around in an endurance test we did, long enough to court again, and long enough to mate again."

The males certainly do their best to survive as long as possible. But, alas, after the second mating, that's it. Not only has the female had time to chew all the way down to his internal organs, but the male has run out of sperm, so he has no incentive to keep going. That's another key point of spider biology. Mammals produce sperm in the same set of tubes used to deliver them, but in spiders these two functions are separate. Male spiders produce sperm in a gland on their abdomen, then they deposit it on their web, and collect and store it in one of their palps. That means that even if the males survived the second mating, they'd have to go through the whole sperm manufacturing process before they could mate again, which, considering the damage they've received, would be unlikely to succeed.

Still, the question remains, why go through such a dramatic process to mate? This is a one-shot approach for the males, and if it fails, their genes never get passed on. The answer is that up to 90 per cent of males never find a female to mate with. The odds of finding a second female are so low that the males are better off throwing everything they've got at one mating, rather than trying to find more females.

And if you're only going to get to have sex once in your life (if you're a redback spider), you'll go to any lengths to make sure it works, no matter how fatal that might be.

This suicidal behaviour isn't the only odd thing about the male redback spider. He is also much smaller than the female, by as much as forty times, yet being small doesn't give him an advantage over larger rivals – quite the contrary. This fact puzzled Dr. Michael Kasumovic (who is now at the University of New South Wales in Australia, but was one of Dr. Andrade's graduate students at the University of Toronto, Scarborough). He found that when small and large males were placed side by side to approach females, the larger males would always beat out the smaller ones. They would just push them out of the way. Yet, in the wild, small males continue to exist.

The key seems to come out of the cues that cause smaller males to develop in the first place. And the word "develop" is important here. Earlier work by Dr. Kasumovic and his colleagues had shown that it's not genetics that leads to smaller males. Rather, when the males could sense, somehow, that there were lots of females around, but not too many males, then they sped up their own development and hatched, on average, a day sooner, when they were still small. When there were more males than females, then the spiders wouldn't hatch until later, when they were larger.

To see if this led to an advantage for the early-emerging spiders, Dr. Kasumovic set up two different conditions. In one, three large and three small males were all released together to find females. As expected, the big males beat out the smaller ones. But if the small males were released a day ahead, as would happen in the wild, then they got to the females first.

Dr. Kasumovic explains, "When there are a number of males all searching for the same female, and they arrive at the female at the same time, there is going to be a benefit of larger size. When there are more males around, it's better for him to take a little bit longer to develop, to make sure that he's a little bit bigger than the average size. So he can actually compete and out-compete these other rival males when he does reach a female's web." But if there are lots of females and fewer males, he's better off rushing to get there first.

It's a curious balance. There are advantages to both approaches, and, depending on the details of the season, each one has a chance of winning. Which just goes to show you: sometimes, size isn't everything.

SOME CRICKETS LIKE IT ROUGH

The cricket equivalent of dinner and a movie before a first kiss is the song the male produces by vibrating his wings. Female crickets like to be serenaded before letting a male mate with them. But according to Dr. Karim Vahed, from the University of Derby in England, there's one species, the alpine bush cricket (*Anonconotus alpinus*), whose males are so virile that they don't bother with such formalities.

If you want to see the alpine bush cricket in action, you'll need to head to the French Alps. There, you'll be looking for a brown, plump cricket, about two and a half centimetres (one inch) long. Other than their size, they look like any other cricket.

What's special about them is their mating behaviour. We've all heard crickets call – it's one of the quintessential sounds of a summer night. The males are advertising their position; the females are attracted to the male, they climb on, and mating ensues. As Dr. Vahed puts it, "The females are on top in the cricket world and copulation occurs." But not for the alpine bush cricket. In this species, the males search through the long grass of the alpine meadows, and when they come across a female, says Dr. Vahed, "they leap on her without any prior warning, wrestle her to the ground, and use a large pair of pincers at the end of their abdomen to maintain a vice-like grip while they begin copulation."

If that wasn't enough, these males are so highly sexed that just eighteen seconds after they've finished mating they're ready to jump on another female. Or, indeed, on any other cricket that's around. They are extremely indiscriminate when it comes to mating and will jump on the back of any cricket that's close by, male or female, young or old. This eighteen-second turnaround is remarkable – the males of other cricket species can take as long as five days to recover and be ready to mate again.

However, there has to be a limit to how many times they mate, or, as Dr. Vahed says, their lives would be "short but happy." In captivity, he's seen males mate as many as seven times in quick succession, and he'd like to study some males in the field to see if they also mate this frequently and rapidly.

That's the male's side of the story, but there's another player here, the female. She comes off the worse for this encounter. Dr. Vahed says, "When you collect females from the field, you'll find that most females have distinct bruising on their abdomen in the region that's first gripped by the male's pincers." Which makes you wonder, why doesn't she resist? The truth is, she does. When mounted, Dr. Vahed says, the female "kicks and bites at the male, and also they have a technique of walking quickly through vegetation to try and brush the male off. In most of these cases, when the female does choose to resist, she actually succeeds in dislodging the male." One theory is that by resisting the male, the female is ensuring that only the strongest, and therefore healthiest crickets are getting to pass on their genes. Only the best of the males get away with the rough behaviour.

As weird as this behaviour is, there's one more fact that just adds to the strange story. These silent males are perfectly capable of chirping. In the lab, Dr. Vahed has isolated males and found that after eighteen days or so, they'll revert to calling for females. Which suggests that there are two different strategies this species can use. When there are plenty of females around, the males just sneak around and jump on unsuspecting females. But when the females are few and far between, it pays to advertise.

Of course, if any male's approach is to sneak up on females, it doesn't make sense to sing. Stealth may be what explains their silence.

The Others

Insects aren't the only species that engage in a real battle of the sexes. In fact, it's found throughout the animal kingdom. And as it does with insects, the battle drives some creatures to extremes...

Genitalia Most Fowl

Did you know that most male birds don't have a penis? For most avian species, sex is a matter of lining up the male's cloaca – that's the small hole on the bird's body where the poop comes out – with the female's cloaca, and passing sperm from the male to the female. But ducks are a spectacular exception, and the Argentine lake duck (*Oxyura vittata*) has the longest penis of any vertebrate in relation to its own body length. The big mystery (beyond why, of all the birds, the duck has a phallus) is why this duck's is so long. Dr. Patricia Brennan, a behavioural ecologist at Yale University, thinks she knows what the answer is. She discovered it by looking where few have dared to go: inside the duck.

As impressive as the male duck's penis is, very few people have ever seen one. For most of its life, the phallus is hidden inside the duck's body, coiled up inside the cloaca. It emerges only when he mounts the female to mate, and then it's inside her body, so you're not going to see it then, either. Dr. Brennan first saw one when she was studying what happens between ducks after copulation is over. As one male dismounted after sex she caught a glimpse of the phallus. "I was completely shocked by this organ," she says, "which is unlike anything I had ever seen before. It's a corkscrew-shaped, whitish, tentacle-like organ with all kinds of bumps and ridges." Its corkscrew shape makes it quite different from most animal penises, and depending on the species of duck, it can be anything from two centimetres long up to forty-five centimetres (three quarters of an inch to almost eighteen inches).

Seeing an organ like this raised all kinds of questions in Dr. Brennan's mind, not least of which was, "Oh my gosh, this is incredible, but what does the female look like inside?" After all, if the male has such an extraordinary phallus, there must be a good reason for it, and that's going to have something to do with the female anatomy.

So, Dr. Brennan got herself a deceased female duck and dissected the reproductive tract. "It was very strange," she says. "It's not a simple tube, like it is in most birds. It's actually a very complicated structure, with several blind-end pouches immediately at the entrance of the cloaca." If a male's phallus ended up in one of those pouches, any sperm he left behind would be wasted, since they couldn't get anywhere. And that's just the first complication for the male. In the female reproductive tract, says Dr. Brennan, "there are series of spirals all the way to the shell gland [where the shell is formed around the egg]. And these spirals are in the opposite direction of the spiralling of the male phallus."

Imagine the plight of the male, then. He mounts a female, and if he's not lucky, his phallus ends up in one of the pouches, and he deposits his sperm, but it just dies. Even if he manages to find the right passage, his phallus will have a hard time fitting. It certainly doesn't look like a system that's intended to maximize fertilization. Quite the opposite. It looks as though the female is trying to prevent herself from getting fertilized.

But as strange as this all seems, it does makes sense, at least from the duck's perspective. When ducks fly to their winter breeding grounds, the females of most species will have already selected a mate. However, there are plenty of other males around who will try to force themselves on the female. "And the females," says Dr. Brennan, "will strongly resist these attempts from the males. They try to get away. They fly away. They try to escape. But oftentimes they can't. In these cases, these sort of rogue males are successful quite often in submitting the females to this forced copulation." This complicated reproductive tract, then, decreases the chances of successful fertilization by these unwanted males. When the females are with their chosen mates, they don't resist, and he's able to wiggle his way to reproductive victory.

What it's led to is another evolutionary form of the genital arms race. It's to the male's advantage to have a long penis that can get all the way into a female's genital tract, and it's to the female's advantage to make the process as difficult as possible. Luckily for the species, it's not a complete stalemate, or the ducks would die out. But it does make for some weird anatomical structures.

Dr. Brennan now wants to train male ducks to mate with artificial vaginas, "and see if we can get an idea of really how this works, and get a window on what the interaction inside of the female might be like." Made out of glass or silicone, the vagina would allow her to see how the phallus uncoils and makes its way through the tract. The things some researchers will do in the name of science!

CANNIBAL FISH

Nature has come up with some pretty effective ways of keeping populations under control. There are predators that eat unwary or weak prey. Food scarcity restricts how many animals survive, or, in some species, limits the number of offspring a generation can support. There are diseases that wipe out a segment of the population, typically the very young and very old. And then there's cannibalism.

Eating members of your own species seems like an odd form of population control. After all, biologically, the whole point of life is to see your own species survive and multiply. Yet, according to Dr. Suzanne Gray, a biologist at McGill University, many fish practise cannibalism, and eggs spawned by others are a common food source. Among those species that take care of their offspring, the fish that's minding the eggs (most often the male) usually eats any that have been fertilized by another male. But in one species alone, as far as we know, the males actually gobble up their own offspring.

This fish species is *Telmatherina sarasinorum*, and it is found in only one lake, on the island of Sulawesi, in Indonesia. It's not a species that cares for its own young. Once the eggs are fertilized, they are left to make it (or not) on their own. The males are the well-dressed partner in this species, with colourful fins and one of five body colours (the two most common are yellow and blue). The females are silver and blend into the background. If you saw them, you'd say they were quintessential tropical fish, with both males and females measuring about five centimetres (two inches) long. Not that you should put them in your aquarium – they're on the endangered species list.

And their own behaviour isn't helping them keep their population levels up. When Dr. Gray first went to Indonesia to study these fish (at the time she was working on her Ph.D. at Simon Fraser University), it wasn't long before she saw the curious cannibalism in action. It all happens when the fish spawn.

Most of the spawning is typical, run-of-the-mill fish behaviour. It starts with the male showing off, as males of all species are prone to do. He swims around the female, displaying his fins, seeing if he can get her interested. If that works, then the female and male both go down to the bottom of the lake, so that the female can lay her eggs. She quivers, lays the eggs, and soon the male heads in to fertilize them. But then, Dr. Gray says, "The male fish will turn around, and it's as if he's inspecting the area where the eggs were just laid. And about 12 per cent of the time, he eats the eggs that were just laid."

There's something special about this 12 per cent of the time. That's because there's one other player in this story, and that's the sneaker male – one who's trying to get in on the action. That's one of the dangers of breeding in the open. One male does all the work, then another one jumps in at the last moment to fertilize the eggs. And, in this species, when it appears that's what's happened, the first male swoops back in to eat the eggs. As Dr. Gray says, "The male that was paired with the female is making some assessment about how related he is to those eggs. If another male comes in and cuckolds the spawning event, it's less likely that the first male is actually going to fertilize those eggs, and so it's to his benefit, then, to eat the eggs."

What benefit? The answer is energy. The male puts a lot of effort into his courtship of the female. All that swimming around and fighting off other males is real work. And these males fight! They'll bite, and nudge, and really work up the fish equivalent of a sweat, enticing their partner to spawn. If another male is getting the benefit of passing along his genes, they're really losing out. It's better for him to eat the eggs and get back some of his energy, since eggs are very nutritious, and then try again to persuade a female to spawn.

Dr. Gray noticed that there's real competition among males to get those little bundles of eggy goodness. The sneaker males quite frequently eat the eggs too, probably because they know that many of them have been fertilized by another. So if the original male gets there first, he's getting an energy burst that otherwise would have gone to his competition. Apparently, it's a fish-eat-fish-egg world in this lake. But why this Indonesian species has evolved this behaviour is still a mystery.

Another mystery is why they don't eat only their own species' eggs. In some follow-up work, Dr. Gray has seen male fish from this species eating the eggs of other, related fish. Maybe they just have a taste for faux caviar?

INFERTILE MONKEYS

Living in a social group has definite benefits. For instance, dividing labour among the group's members means more work gets done, and there are lots of you around to help raise the kids. You've also got some extra protection against outside enemies. And, last but not least, there's a good chance there will be someone around to keep you warm at night. But you've got to play by the rules if you want to take advantage of these benefits, and for the common marmoset monkey (*Callithrix jacchus*), one of the rules is that only the dominant female gets to reproduce. All the other females have to settle for a chaste life. Biology has provided a helping hand to make this easier for the subordinate females. As Dr. Wendy Saltzman, a behavioural biologist at the University of California, Riverside, has discovered, these monkeys have evolved a way to make their celibacy bearable.

Common marmosets are New World monkeys that are found in Brazil. They're about the size of a squirrel and have tufts of white hair sticking out around their ears. In the wild, they live in groups of between five and fifteen, mostly made up of animals that are closely related to each other. The females usually give birth to twins, and these babies are relatively large, proportional to their adult size, compared to the young of other primates. Another unusual feature of the common marmoset is that they can breed twice in a year. Put this together and we're talking about a species with a high reproductive rate and young that are biologically expensive to raise. This probably explains why only the dominant female gets to breed. She needs the help of all her sisters and other daughters if she's going to care for four kids every year.

What's interesting about the subordinate females is that, not only do they not engage in sexual activity with males, but they're also biochemically prevented from wanting to. "In the lab," says Dr. Saltzman, "we know that as soon as a female becomes subordinate in a social group, she stops ovulating and that means she's infertile. Her pituitary hormones drop down and those would normally stimulate the ovaries. Clearly, something's changing in her brain that basically switches off the reproductive system in response to social cues."

In other monkey species, subordinate females often breed less often than the dominant female, but that response is usually stress-induced. The dominant female bullies the other monkeys, which discourages their breeding. But in this species, there's no evidence of stress levels being raised among the subordinates. Simply, once they realize they're subordinate, they switch off the sex hormones. "What's interesting," says Dr. Saltzman, "is that there can be other females who are also living with what would seem to be a dominant female, but if they do not consider themselves subordinate, then, as far as we can tell, they do not show this reproductive suppression. There's something about the perception of their own social status."

And the sex hormones can be switched on again. When Dr. Saltzman moved female monkeys from group to group, she found that they would engage in sex and produce young if they found themselves to be the dominant female. Conversely, if they became a subordinate, then even a formerly dominant female would drop her reproductive hormones. Says Dr. Saltzman, "It's an amazing switch that turns fertility off or on in response to the social environment."

There has to be a good reason behind this switch, and Dr. Saltzman thinks she knows what it is. "When subordinate females do breed, for whatever reason, their infants are likely to be killed. From what's been seen, especially in the field, it's clear that this is being done by the other breeding female, by the dominant female." If this is the case, it's not in a female's interest to put energy into producing offspring when they're only going to be killed later. Remember too that the animals in the group are usually all related. So raising another marmoset's offspring means that at least some of the genes of the subordinates are being passed on, which is better than nothing. And, from the point of view of the group, there's a better chance of the dominant monkey's offspring surviving to adulthood if the whole troop is helping to raise them.

So, while only one female gets to reproduce, at least the family's genes are being passed on, and, evolutionarily speaking, that's really what life is all about.

Meerkat Infanticide

If you watch much nature television, you're sure to have seen meerkats (*Suricata suricatta*). Who wouldn't want to watch them? They're very appealing animals. After all, they seem to be cooperative to an extreme, always looking out for one another. You've probably seen pictures of the meerkat's most famous pose, up on its hind legs, standing erect as a scout while the rest of the family is out playing or hunting for food in the Kalahari Desert of Botswana and South Africa. They also share the raising of their offspring, even to the point that females without their own young produce milk and nurse the infants in the troop. But there's a dark side to meerkat life, one that's been uncovered by Dr. Andrew Young, now a researcher at the University of Exeter in England. It turns out that females are engaged in a bloody battle to ensure the supremacy of their own offspring.

There's a great deal at stake here. Out in the desert, resources are limited, and there's only so much food to go around. This means there has to be a limit on the number of babies the troop can support, which in turn leads to a limit on who gets to reproduce. Typically, only the dominant females in the group breed, and the subordinate females help raise the kids. This isn't such a bad deal.

All the females in the group are related, so while the subordinate ones might not be raising their own offspring, at least the animals are their genetic kin. But Dr. Young's work showed there's more going on than just this.

"What we've uncovered," he says, "is that there's actually a multi-party power struggle going on here. Whenever any individual female gets the chance to breed, even if she's a subordinate, she'll use all the tactics at her disposal to maximize the chances that her own pups will survive. And one of the best ways to do that is by killing off the offspring of other females."

It's as brutal as it sounds. Dr. Young watched pregnant females go underground to give birth. Then, the next morning, he'd see another female, who was always also a pregnant female, "drag the newly born pups to the surface and kill them off." It gets worse. "It can become quite messy," he adds, "and then she'll typically eat the pups. But the remarkable thing is that the mother doesn't seem to react in any obvious way towards the killer of her offspring."

His hypothesis is that these subordinate females are trying to increase the chance that their own offspring will survive on the limited resources of the colony. How she gets access to the pups isn't clear, as it's all happening out of sight underground, but Dr. Young assumes that the mother puts up a fight there, and it's only after she's lost control of the young that she also loses interest. There's no evidence of any revenge taken on the subordinate who stole the kids. "This may ultimately make sense because these females are all related," Dr. Young says. "So, if one female kills off another's pups to feed her own, the female that's had her pups killed will still benefit indirectly from the success of her sister, of her sister's litter. So it's not quite the genetic cost that it would be if the females were unrelated." Keeping it in the family, as it were.

This whole situation is very unusual. In some species, dominant animals might kill off the offspring of subordinate animals, but it's rare for the subordinate even to get a chance to kill the kids of her superior. But the meerkat mother has to leave her young to search for food, so they're at risk of being taken by other animals.

It's also unusual for subordinate females to get the chance to get pregnant. Most of the males around are the brothers of the females in the colony, and biology tends to limit reproduction among siblings. Natural selection has weeded out incestuous behaviour over time, since it is more likely to allow bad genes to stay in the population, and genetic diversity is important for a species' survival. Dr. Young says, "For a large part of the subordinate's life, it may simply be that they don't have access to mates. But then there are situations where, for example, you may have an immigrant male join the group, or the female will have the opportunity to mate with passing males, and subordinate females do become pregnant. And that's when these types of problems emerge."

Most of the time, then, the situation is "hunky-dory," as Dr. Young puts it. Only the dominant female meerkat gets pregnant, and the colony as a whole works to make sure the offspring survive and grow up. It's only about a quarter of the time that a subordinate female will be pregnant at the same time as a dominant, and then about half of her offspring will be killed. This amounts to about one in eight offspring of subordinates lost to infanticide, which, while significant, explains why the colonies don't go extinct because of this behaviour.

And if you're wondering where the males are in all of this, they're not involved. They have their own dominance structure, but they've not been seen getting involved in infanticide, unlike the males of other species, like lions.

So the next time you're watching a nature special on the happy, carefree meerkats, remember that there's more going on under the surface than you can see.

Hyena Hormones

The spotted hyena (*Crocuta crocuta*), also known as the laughing hyena, has always been seen as a strange creature. First of all, there's that uncanny call – the hyena's laugh. And, before scientists such as Dr. Kay Holecamp from Michigan State University investigated, many people thought the hyena was a hermaphrodite. This isn't as unreasonable an assumption as you might think. Hyenas do come in both sexes, but the females are larger and more aggressive than the males and even have a penis and a scrotum. Well, not exactly, but they have a prominent penis-like clitoris surrounding their vagina (and, yes, they get erections) and their labia have fused to look like a scrotum. That isn't all. They even produce testosterone, the male sex hormone, which they use to help maintain the hyena's social structure. So much for sexual stereotyping.

The curious story of the spotted hyena really starts with their social structure. No other carnivore lives in as strict a hierarchy. Every animal knows its place in the hierarchy, and where each individual fits determines how much food it gets from a kill. The females rule the roost when it comes to access to food. Socially, the females outrank all the males, and a female's rank comes from her mother. It's strictly a matriarchal society, where the females have literally taken on the characteristics of the male.

Now, back to that weird pseudo-penis. How, you're wondering, do they mate? Well, according to Dr. Holecamp, "When they copulate, it's tremendously funny to watch because the male actually has to insert his penis into her penis-like structure, which is as long and just as tubelike as his. She urinates and copulates and gives birth with this strange organ, but it points forward, so he's sort of got to crouch down behind her and dance around back there until he can achieve intromission, and it's very difficult." It's true. She's not making it up. It does make for difficult births, though.

That's what we can see going on, but there's lots else happening that can't be witnessed. When the female is pregnant she gives the fetus a dose of testosterone during its early development. When the offspring is female, Dr. Holecamp says, this leads to the aggression and large size seen in females, and probably the pseudo-penis and scrotum, as well. That would fit with what's known in human development when too much testosterone is present in early fetal development and daughters are born looking like sons.

Interestingly, from a scientific perspective, just how much of these male hormones a female delivers depends on her social rank. The more dominant she is, the more of the male hormones she secretes. Dr. Holecamp believes the key here is how this affects the offspring's chances of survival. After a hyena kill (and they're very good hunters), all the animals from the group gather round to feed. Those who are higher on the social scale get to eat first and most. If you want your offspring to get food, they're going to have to fight their way ahead of everyone else's young. Give them more hormones, and they'll be better at getting in there and eating. That's true whether or not they are female offspring.

There's a secondary effect, too. As mentioned, sex among hyenas is tricky, to say the least – Dr. Holecamp says, "Imagine two animals of this sort having sex through two different penises. It's a very tricky undertaking" – and when they're under six months old and not yet fertile, they get to practise. The more sex hormones they've been exposed to, the more willing they are to learn how to mate successfully.

The fact that only the most dominant females produce a lot of testosterone for their offspring says there must be a cost to all this. More male hormones may make conception more difficult, and lower-ranked females may need to keep their fertility levels as high as possible. Which is lucky for the offspring – otherwise there might be a hormonal arms race, and who knows what the pseudo-penis might come to look like if that happened.

Speedy Sperm

When it comes to mating, males have an astonishing array of tools that help them compete with other males in attracting females. There's the peacock's magnificent tail, the stag's regal antlers, the lion's luxurious mane, and so on. These tools work well when mating is about choosing a single suitor. But what happens when a female doesn't chose just one mate, but many? Well, it turns out there's still a way for the males to compete. Dr. Sigal Balshine, a behavioural ecologist at McMaster University in Hamilton, Ontario, has discovered that, in some species of fish, becoming a better swimmer is the key to sexual success. But that swimming might not be quite what you'd expect.

The fish are cichlids. These tropical fish come in all shapes and sizes, from all over the world, and are popular aquarium pets because of their brilliant colours and the fact that they take care of their young. One of the best places to find these creatures is Lake Tanganyika, one of the African Great Lakes, where there are more than two hundred described species. As Dr. Balshine says, "There are so many different species of cichlid in this one lake it boggles the mind. It's sort of like going down into a coral reef without the coral, but the fish are equally diverse and absolutely magnificent, all doing very different kinds of fascinating behaviours."

A place like this is a behavioural ecologist's dream come true. Side by side, you have species of fish that are closely related to one another but have completely different lifestyles, including sexual habits. Alongside one another you might have a species that's monogamous, where the male and female stay together to mate and to raise their offspring, and another where the female mates with several males at once and only the male tends the young.

We're talking fish here, though, so mating isn't quite the same as it is for mammals. These fish have a couple of different approaches. Sometimes, the female lays the eggs on the lake bottom and the male comes in and releases his sperm on top of them, a common practice among all fish. In other cichlid species, the female lays an egg in a male's nest, picks it up and puts it in her mouth, then approaches the male, who deposits his sperm directly into her mouth. This second approach is quite intricate. In Dr. Balshine's words, "He does this sort of courtship dance for her. And as he's courting, she nuzzles his anal fins, which have egg dummies on them, the very same shape and size as her own eggs. If you like, she's tricked into thinking these are eggs. She nuzzles these egg dummies and he squirts sperm into her mouth." Before you wonder why a cichlid might do this, you should know that many species of cichlids actually hold the eggs in their mouth until the babies hatch and are released into the water. So it's not that odd for fertilization to happen in the mouth, too.

Back to the question of competition. Whether the eggs are deposited on the lake bottom or in the mouth, the females of some species will have them fertilized by more than one male – which is going to lead to competition. After all, the first fish to get to the eggs has the best chance of fertilizing most of them. If his sperm get to the eggs first, he wins the reproductive competition. Based on this, you'd expect the sperm from promiscuous species to be longer, faster, and possibly bigger than those from species in which the males aren't competing for access to the eggs.

Dr. Balshine knew that Lake Tanganyika would be the perfect place to test this hypothesis out. She collected males from different species, stripped them of their sperm, and compared their appearance and swimming skills. When she looked at the sperm through a microscope with an attached high-speed camera (to catch how fast the sperm were moving), she got a more dramatic result than she'd expected. First of all, the sperm from promiscuous fish were larger and longer than the sperm of monogamous species. And they were really fast, at least for sperm, swimming between 50 and 100 microns per second. That's not anything like how fast we can move, but these sperm are really small, so that's a high speed for them. Even more importantly, though, these sperm were swimming for as long as ten to fifteen minutes. That's endurance, and it was missing in the sperm of monogamous males. Of course, the monogamous males don't need big, speedy, strong sperm, so they would opt for smaller and slower sperm. They take less energy to produce, meaning the male doesn't have to spend as much time and effort finding food as his promiscuous cousins.

What this tells us is that competition between males for access to females is fierce in the promiscuous species of fish. And that competition has led to subtle but important changes in the fish's physiology. While it hasn't been as widely studied in other animals, this sperm competition may well be widespread in nature. It turns out you can learn a lot about a species by looking at the male's sperm.

2

IMPRESSING THE OTHER SEX

Birds do it

❖

Finding a mate is very important; otherwise, how are you going to breed? But in the animal world, the competition is fierce, so you have to attract attention. And of all the animals, birds have some of the most visible and advanced approaches to picking the perfect partner.

Zebra Finch IQ

It's probably no surprise to learn that female songbirds like a male who can sing a nice little ditty. It's one of the main reasons male birds warble at all. But what makes a male song so attractive to a female? Are they belting out the equivalent of "Hunk'a Hunk'a Burnin' Love," or are they just proving they have a good set of lungs? Neeltje Boogert, a Ph.D. candidate at McGill University, has been trying to decode the secret of songbird success by looking at the zebra finch (*Taeniopygia guttata*) and has found that the song isn't a sign of beauty or strength or romance. No, for the zebra finch, it's a sign of a healthy brain.

The zebra finch has become a favourite of researchers the world over. It's native to Australia but has been in North America for a long time, mostly as a pet. It is a small bird, weighing only about fifteen grams (half an ounce), but you'd probably recognize one if you saw it. The males have bright orange cheeks, and a bright orange beak, and zebra stripes across the chest. And while their songs don't sound particularly melodious to a human, female finches are adept at telling them apart.

The differences among songs are extremely subtle to the human ear, but with the proper equipment, they are clear. Some songs are longer than others. Some contain more or fewer notes. Depending on how many notes there are, and the length, researchers give each one a complexity score. Each male has a unique song, which he repeats over and over. It's his signature, if you will. Interestingly, the males learn their songs from their fathers. But they're not perfect copies. Two sons of the same bird will have similar songs, but they'll never be exactly identical. The question, then, is twofold. First, what does a female finch look for in a song? And second, why?

The answer to the first question is well established. Females prefer the more complicated songs. Which then leads to the second question. As Ms. Boogert puts it, "If a male sings a really pretty song and the female chooses this male for its pretty song, what does she get in return, in addition to the pretty song?"

To study this, Ms. Boogert gave the finches a task. Imagine this set-up: a wooden block drilled with holes inside which food is hidden. The bird has to find the food in the holes. Except that's not much of a challenge for the finches. So, Ms. Boogert put cardboard lids over the holes, which the birds had to flip out of the way if they wanted to get the treat.

There were four different trials in all, and Ms. Boogert kept score. Some finches got it correct right away. Other birds never did figure out how to lift up the lids and grab the food. In between, there was a range of times and speeds, depending on the individual finch's abilities. Ms. Boogert says, "When we looked at the pattern of how this intelligence score related to their complexity score, we found that the males with the most complex songs were also the fastest at solving this problem."

From what we know about zebra finches, and birds in general, this leads to an interesting combination of ideas. Birds have a special area of the brain that's responsible for creating song. The more complicated the song is, the larger that area is. "And," says Ms. Boogert, "the really interesting thing is that the volume of this brain area for song actually correlates with the volume of the overall brain. So if you have a big brain for song, you have an overall bigger brain. And this makes sense, then, for what I found out. The more complex singers with the bigger song areas also have a bigger brain that allows them to be more intelligent."

And who wouldn't want an intelligent mate? Especially if you're a bird, and the intelligence allows your mate to excel at finding food. Not only that, but this big brain is heritable. So if Dad's got a big brain, so will his offspring. And, in the world of evolution, there is nothing more important than having offspring that are going to be able to feed themselves well and pass your genes on to another generation.

Maybe there's something for us humans to learn in all of this. Guys, if you want to get the girl, it's the size of your vocabulary that's really going to impress her!

BOWERBIRD BULLIES

Keeping up with the Joneses can be a tiring business. They get a new car. You get a new car. They paint their house. You paint yours. They put up a giant Hallowe'en display. You go over in the middle of the night and knock it down. Okay, maybe not that last one, but, face it, you've probably been tempted. Who wants to be the one person in the neighbourhood with a shabby home? If you were a spotted bowerbird (*Chlamydera maculata*), that temptation would lead to action. According to Dr. Joah Madden, an animal psychologist at the University of Exeter in England, male bowerbirds will trash their neighbours' bowers. And it seems this behaviour is about more than just keeping up appearances. It may have something to do with their reproductive success.

Bowers are a central part of the breeding life of bowerbirds. They are structures, often extremely complex, used to attract the attention of females. Some species make covered shelters so complex that, according to Dr. Madden, "the first European explorers in New Guinea thought they were playhouses." Other species, like the ones Dr. Madden studies, build less complex bowers, called avenues, with walls along two sides, about forty centimetres (fifteen inches) long and thirty centimetres (twelve inches) high.

Around this avenue is the most remarkable part of the bower. With a fastidiousness that would make the fussiest housekeeper proud, the male removes any debris and replaces it with yellow straw. On this, he displays any colourful objects he can find – shells, flowers, feathers, even human junk. If it shines and he hasn't yet finished decorating, he'll use it.

During breeding season, the females will fly around the different bowers in her neck of the jungle, looking for the one she likes the most. Once she's seen them all, she'll go back to her favourite, and its builder is who gets to father her offspring. "What seems to be the case is that many females all agree on which male is the most attractive," says Dr. Madden. "So some males get many copulations, and some get none at all. And it all seems to depend on some component of their bower."

Having a good-looking bower, then, is critical for success. And here's where the raiding comes in. As Dr. Madden puts it, "It's a cut-and-thrust world out there in nature, so if your next-door neighbour is destroyed, if he's out of the game, if his bower is just a pile of sticks, then you're certainly going to do far better with the ladies." So that's what happens. A male will fly to his neighbour's bower and, says Dr. Madden, "grab at the base of the wall sticks, and he'll wobble his head from side to side and generally just rip the whole structure up until it's flat on the ground." Sometimes he'll even steal the best stuff and take it back to his own bower. Not only is he destroying the neighbour, he's improving his own lot at the same time!

Location has a lot to do with who's going to get attacked. Dr. Madden has figured out that the bird most likely to trash your bower is the one next door. And don't think that being related to him is any help. There's no family favouritism in this behaviour. No, the neighbour is always at risk of being the victim of an attack. Which is a little odd, since relatives share some of their genes, and you'd think this would mean that each would want the other to succeed. But not in the case of the bowerbirds. It seems that everyone's watching his next-door neighbours. That's no mean feat. These birds often live one or two kilometres apart (half a mile to just over a mile), so there's plenty of flying back and forth to look at the neighbour's set-up.

Recent work by Dr. Madden has shown that the amount of raiding in an area eventually drops right off. What seems to be going on is that there's a hierarchy in the population. The dominant males have the best bowers, and so on down the rankings. It's only when a male tries to overstep his position, by showing off with extra decorations, that he gets attacked. That's how Dr. Madden did his work. He'd go in, add extra stuff to one male's bower, and watch what happened. Sure enough, that would prompt a raid, and a return to the status quo.

The males themselves seem to know this, too. When Dr. Madden put extra decorations around where males could go and pick them up, they were reluctant to over-decorate their bowers. "They only took as many as they naturally would have anyway," he says. "And when I gave them the opportunity of having a load of extra decorations on the bower, they actively removed them. They seemed to know their place."

The birds need the power to back up their decorating, or, as Dr. Madden puts it, "It's almost like the Mafia. If you're a particularly fit male and you've got good resources, then you flaunt your wealth. You show off your diamond rings, your fast cars, and you reap the attendant rewards of beautiful women, of course. But if you don't have the muscle to back that up, then very soon, you're going to be gunned down."

And the neighbours are certainly paying attention. When Dr. Madden went in and changed bowers, within ten days the neighbours would be by to raid. Which just shows you: keeping up with the Joneses is good, but don't push it, not unless you can back it up with some muscle.

BOWERBIRDS COME ON STRONG

Disco may be dead for human culture, but for the satin bowerbird (*Ptilonorhynchus violaceus*), it might be the soundtrack of its life. This avian John Travolta, circa 1977, needs a hot bachelor pad, a smooth, satin plumage suit in blue, and all the right moves if he wants to land a female bowerbird. He'll lure her to his artfully constructed bower and demonstrate his dancing talents. If she's impressed by his smooth moves, she'll stick around. But if he comes on too strong, she just might leave. Dr. Seth Coleman, an animal behaviour researcher at Gonzaga University in Spokane, Washington, has explored how subtle the line is that the male bowerbird mustn't cross, if he doesn't want to move from stud to dud.

Step one for the bowerbird that wants to get his girl is the building of the bower. The satin bowerbird builds an avenue-style bower. According to Dr. Coleman, "It's a descriptive term because they quite literally look like an avenue or a hallway of sticks, maybe a foot long, with really thin sticks in each wall, and there's an avenue between the walls of the bower that the female crouches in during courtship." Around this avenue, the bird lays out a collection of bright, shiny objects to attract a female.

A beautiful pad isn't enough, though. If this male wants to land a mate, he has to show off his moves. After luring a female into the centre of his bower, it's time to dance. For the satin bowerbird, this is a complicated combination of moves. First he'll fluff up his feathers. Then he'll flip up his wings, make a loud buzzing sound, and start to run back and forth in front of the female.

It's a fairly aggressive dance. And if he's trying to impress a young female, well, too much dancing and she'll be scared off. The younger female bowerbirds are skittish, and a smart male knows he has to avoid coming on too strong if he wants to win her heart. For the young females, the secret to success is the quality of the pad, not so much the dance. If the bower really shines and is beautifully decorated, then she's going to stick around. Mature females are a different story. They aren't as interested in the fripperies he has so carefully displayed for her. As long as he can build a moderately decent bower, then, really, it's the male's physical ability that counts. Or, as Dr. Coleman puts it, "I've already assessed the real estate. Now let's see if you sing and dance."

It's the smart male that can tell which of his talents he needs to emphasize. In one experiment, where the research team set up robot female bowerbirds, they found that smart males adjusted their dance displays to tone them down or pump up the action, depending on the responses they got from the robot birds. In fact, among the bowerbirds, the most attractive males are the ones who are better at making these judgment calls. The least attractive have a tendency to scare off the younger females by coming on too strong.

This all shows that males need to have a variety of strategies if (please excuse the pun) they want to pick up (and make) chicks. What works with a young female bowerbird isn't going to be successful if he's after a more mature mate, and vice versa. And it says something about the female, too. Her tastes clearly change over her lifetime. For a young female bowerbird, it's what the male owns that makes him a good catch. Then, later in life, it's how he shakes his tail feathers that makes her want to mate.

How this translates to other species, including our own, is something best left up to the reader to ponder.

Godwit Vacations

The Icelandic black-tailed godwit (*Limosa limosa islandica*) is an admirable bird. It has a stable family life, practises monogamy, and usually sticks with its mate for life. Sounds boring? Read on.

Since Iceland is a tad chilly in wintertime, godwits spend the off-season in warmer climes: the south of England, the coasts of France, and even sunny Spain. All sensible choices. And, in another sensible move, but one that's quite remarkable, the godwit has developed a strategy for keeping its relationship with its "significant other" from growing stale – the mated pairs vacation separately. Dr. Jennifer Gill, an ornithologist at the University of East Anglia in England, has discovered that when the birds begin their annual migration south, they go their separate ways and don't reunite until the following spring.

If you're ever visiting Iceland and want to see these remarkable birds, head down to the seashore. They are wading birds, with long legs and long beaks, and they spend their summers on the mud flats of estuaries, feeding on small insects and other invertebrates. They stand about twenty centimetres (eight inches) tall, and are a reddish brown colour in the summer and grey in the winter.

The discovery that the pairs split up for the summer was a surprise to the researchers. Dr. Gill and her colleagues have been studying godwits for a number of years and, over time, have banded hundreds of individual birds that they then track on their migrations, thanks to the help of European birdwatchers. "The remarkable thing," says Dr. Gill, "is that we found that pairs of godwits we'd been studying in Iceland go off to completely different winter locations. But then the following year, they somehow manage to arrive back in Iceland within just a two- to three-day window of one another."

The separate locations are quite widespread. In one pair Dr. Gill studied, the male spent the winter in Ireland, while the female was in the west of France. Yet somehow they arrived back at their breeding ground just a couple of days apart. It was almost as if they'd phoned each other to coordinate their return. Except, as far as Dr. Gill can tell, the birds don't carry phones. This goes on year after year. The birds spend three months in the breeding area, then nine months apart. Considering that these birds are known to live more than twenty-five years, that's a lot of time in a long-distance relationship.

This timing of their return is vital for each couple. Dr. Gill was able to show that if the female arrived as few as eight days before her mate, she would go off and find another male to pair up with. There's a cost to this, though. Females who divorce have fewer offspring that season. It's not that the new male isn't a good partner, but when two birds set up house for the first time, they have to find a nesting site and build a whole new nest. This takes time, time that the established pairs don't need to spend so they lay eggs sooner. The sooner they lay, the more offspring they are able to have. So, in the end, divorce happens only if it's absolutely necessary.

This means it's to the advantage of both birds to time their arrivals as closely as possible. While no one's really sure exactly how they do this, there are some hypotheses. One, suggested by Dr. Gill, says the timing is dependent on the quality of the wintering sites. If both birds are on good sites, then they'll be in good shape to come back to the breeding ground at the same time. If that's their first breeding season, then they'll meet a member of the opposite sex and pair up. Then, the following winter, the same cues at the wintering site will send them back to the breeding ground. Once again, they'll meet their partner. "But still," says Dr. Gill, "a two- to three-day window is quite remarkable."

Remarkable as it is, it's highly successful. Dr. Gill estimates that 90 per cent of the birds make it back close enough together to avoid divorce. What isn't known yet is whether a bird would go back to an earlier partner in the future if that bird's timing were better. Which is a rather romantic notion.

It's possible that the godwits aren't alone in going their separate ways over the winter. Many species of birds stay paired during their migrations, but not all. In those species where the males and females leave independently, it's entirely possible they're heading to different locations. Nothing like spending time apart to keep a relationship strong.

Auklets and Aphrodisiacs

Another Saturday night, and you're heading out on a hot date. You've showered, brushed your teeth, now you're deciding what kind of perfume or cologne to wear. How about reaching for the bottle of bug spray? Okay, maybe citronella doesn't say passion to you, but that's exactly what it means to the crested auklet (*Aethia cristatella*). This Alaskan seabird thinks that the smell of citron is beautiful, so much so that it produces the scent itself. Dr. Hector Douglas, an ornithologist from the University of Alaska, Fairbanks, was the first to work out why lemon means love for this small bird.

If you've ever visited the coast of Alaska, you may have come across these birds. They're relatively common there, although only during the breeding season, which runs from May to August. The rest of the year, these birds spend their lives at sea in colonies, which can contain hundreds of birds. Standing about twenty-three centimetres (nine inches) tall, males and females look alike. Most of the body is grey or black, but there are a few distinctive features. Dr. Douglas says, "They have a very bright orange bill, a long crest of feathers that sprouts from their forehead and droops down over that bill. And they emit this citrus-like scent during the breeding season."

It was this citrus scent that attracted Dr. Douglas's attention. And it's hard to miss. "You can smell it quite well on days when it's relatively warm and there's not much of a breeze," he says, "from quite a considerable distance, up to a kilometre [over half a mile]. So the smell can be fairly strong."

The first issue he looked at was the reason for the birds to smell this way. Most seabird colonies smell, but usually it's pretty repulsive. For Dr. Douglas, the key was the citrus scent. In other circumstances, as anyone who's been camping knows, citrus smells are used to repel insects. And, in these arctic colonies, there are plenty of insects to worry about. Ticks are the big problem, but there are a lot of lice, as well. If this citrus scent could either keep away or harm ticks, that would explain why the birds smell lemony.

Dr. Douglas decided to experiment. "I put concentrations of these chemicals on filter paper in a Petri dish," he says, "and I measured the movement of ticks over that graph paper. What I noticed is that, over time, the ticks had less and less motor control, and they were slowing down and losing control over their appendages." This is important. One of the key characteristics of ticks is that it takes them up to twenty-four hours to settle down and start to bore a hole into the animal they are going to lay their eggs in. If the chosen host produces a chemical that slows the tick down, it'll never set up home, and the animal will not get bitten. As ticks carry Lyme disease, this is a real advantage. This also starts to explain the romantic aspect, because, as Dr. Douglas puts it, "If you can anoint your mate with something that is going repel ticks, you've got a better chance that that mate is going to have all of its wits and good foraging skills about it, to help raise your chick that year."

This hypothesis is supported by a behaviour of the auklets that has been observed for years. When a male and female auklet court, they intertwine their necks and rub their heads against one another's napes. Knowing that the birds produce a citrus scent, and that this chemical is capable of stopping ticks, the question for Dr. Douglas was whether this behaviour was related to the birds' citrus scent.

Sure enough, he found specialized feathers on the neck that he calls "wick feathers." These specialized structures are where the citronella chemical is secreted. Then, during courtship, says Dr. Douglas, "A male often will solicit anointment behaviour by adopting a horizontal posture, with the head down, and making a little choking call. It solicits the attention of a female, which rubs its bill over those wick feathers, and up over the male's neck and head. This is reciprocated between males and females."

Not only does this courtship behaviour spread the citronella, but, just like applying sunscreen to that hard-to-reach spot in the middle of your back, these birds get better coverage when another bird spreads the stuff on them. It's a great bonding ritual, with a physical benefit, too!

The smell repels insects, but it acts as an attractant for the birds. When Dr. Douglas spread the chemical on foam blocks disguised as rocks in the colonies, the birds would come over, rub up and down, and even push their nostrils up close and personal with the chemical dispenser. The more chemical was present, the stronger the attraction to the birds.

Now, if the chemical industry could just come up with a bug repellent that smells good to us.

THE SOUND OF TWO WINGS CLAPPING

How many times have you been at a wedding and been obliged to do that dance where you have to flap your arms like a bird and clap your hands? Probably more than once, and you'll be happy to know you're not alone. There's a bird that does the same thing. In its case, it's all about getting a mate.

The birds are part of a group called manakins, a family of small, colourful birds native to Central America. The males of one species (the white-collared manakin, *Manacus candei*) have developed a unique mating performance. It's a combination of dance and clapping that would make any avant-garde choreographer green with envy. Dr. Kim Bostwick, curator of birds and mammals at the Cornell University Museum of Vertebrates, has figured out how the manakins put a snap in their dance.

To see these birds boogie, head to the Costa Rican rain forest. Look carefully for an area of the forest where the ground has been cleared of leaves. This leafless ground is where the male manakins display. Wait long enough, and, says Dr. Bostwick, "What you would see would be a bird equivalent of popcorn. They jump around in between little saplings at an extremely rapid speed, too fast to follow with your eye or even understand what's happening. And while they're bopping around between little saplings on the forest floor, they're making these loud snapping sounds."

These snapping noises are like the crackle of fireworks, or the bang of those firing caps that lots of boys play with. Dr. Bostwick refers to this sound as the "simple snap," and it's made by the birds as they flit from sapling to sapling, landing briefly on the ground each time as they go. But this isn't the only noise these birds can make. As Dr. Bostwick says, "There is a second sound. And this one isn't the popcorn-on-the-ground noise. This is made while the bird is perched up in the tree, and it's a loud, rolling crack. It sounds like one sound, but if you could see it acoustically, it's actually eight rapidly delivered pulses."

Firecrackers and thunder cracks, both produced by a bird that's just a small songbird, no larger than eleven centimetres (half an inch) long. For more than a decade, how they made these noises was a big mystery. Dr. Bostwick studied the muscles of the birds for five years to find an answer, without any luck. Then a new tool came along – the high-speed camera – and at last she was able to solve the puzzle.

"With the advent of high-speed video," she says, "I was able to capture these guys displaying, and it turns out what they are doing is they are leaping off a sapling and whacking the backs of their wings together above their back." Think of it as jumping in the air while clapping your hands together behind your back. It's not exactly an easy trick for most birds, as their wings don't reach that far, but the manakins seem to have it down to an art.

That's how they make the simple snap, the sound they make when on the ground. But when they're perched in a tree, it's too hard for them to clap behind their backs eight or nine times in rapid succession. To make the rolling crack, the birds stay sitting and slap their wings forward against each other rapidly. How rapidly? "The pulses are coming so fast," says Dr. Bostwick, "that the birds have to move their wings as fast as a hummingbird would move its wings, to produce the sound that we hear."

There's one question remaining, and that's which parts of the wings are slapped together to make the noise. That's still not entirely clear, but Dr. Bostwick thinks it's probably the part of the feather called the rachis. The rachis is the central spine of the feather, and in these birds, it's swollen right around where the wings would hit. She suggests we're hearing the vibration of the rachae. But without more evidence, she can't be sure.

Between the swelling of the rachae and the changed musculature and bones that allow these birds to touch their wings behind their back, the bird's ability to fly has to be compromised. While studies haven't yet been done to see what the cost to the bird is, with wings so finely tuned, there must be some price they pay. All in the name of love. Males face stiff competition to attract females and will use any tricks they can to get attention.

The repertoire is usually confined to complicated songs and bright, flashy plumage, but in this bird, it includes noises made with their wings. It's all about being better, louder, and more impressive than your neighbours. The bigger the bang, the better the attention.

Better work on that chicken dance for the next wedding you're invited to.

Bees (and Other Insects) Do It

They might seem small, but it turns out that insects are just as choosy about their potential mates as any other creature. They'll even offer up gifts, if it makes them more attractive. And sometimes, the males will do anything, even change their colour, if it'll help them mate.

Hot Insect Sex

Just when we think we have a handle on sex in the natural world, along comes another species to show us just how much we have to learn. The diminutive ambush bug (*Phymata americana*) is one such species. On the surface, this native Canadian insect seems particularly unremarkable, even though it's the surface of the bug itself that has proved most interesting. While many male animals use their fur or feathers or scales or skin to advertise just how hot they are for females, ambush bugs have a whole different kind of heat in mind. For them, sex is solar-powered, as Dr. David Punzalan, an evolutionary biologist from the University of Ottawa, has discovered.

If you've travelled anywhere in southern Canada or the northern United States, you have probably come across ambush bugs, although you might not have recognized them. They're not very large, typically less than one centimetre (a third of an inch) long, and they're usually either yellow and brown or yellow and black. This colouring allows them to blend in with the flowers where they normally sit, and where they perform their ambushes. Because that's what these insects do. Unseen inside the flower, they attack prey when it comes in to feed. As Dr. Punzalan explains, "They have these pretty remarkable-looking raptorial forelimbs, like a mantis claw. And they reach out and grab unsuspecting insects by the tongue, or by the antenna, and they jab them with their piercing mouth parts, inject a bunch of paralyzing enzymes, suck the juice out of them, and discard the empty carcass." Isn't the life of an insect fun?

The discovery of hot sex among ambush bugs was something of an accident. While Dr. Punzalan was out walking, he spotted some of these bugs on flowers, and also noticed that males had darker bodies and heads than the females. Not only that, but there were variations among males, too. "Some males are a lot darker than other males," says Dr. Punzalan. "Their heads are completely black in some cases, while in others they're a little bit paler, and on the sides of the thorax there's a bit of a variation as well. The females completely lack some of these traits."

Being a smart researcher, this difference, both between particular males and between males and females in general, piqued Dr. Punzalan's curiosity. This pattern, in which males and females look different, is usually associated with some kind of signalling. Think of the peacock's tail. Displaying it is the male's way of advertising his health to the female. So, Dr. Punzalan took some of these bugs back to his lab to see if the females had a preference for one colour combination over another.

The answer was no. Dark or light, in the lab, the females didn't like one male better than another. This only made Dr. Punzalan more curious, because in the field, it did look as if darker males were more successful, at least in terms of reproduction, than the lighter males. Something else, other than picky females, had to be a factor. That's where heat came in.

"We hypothesized," says Dr. Punzalan, "that these dark patches were solar panels that allow males to get up and active, like warming up the car in advance. And that this is where they experience a real advantage." It's a relatively straightforward idea. Insects are cold-blooded. They can't move around much until they are warm enough. In this species, the females stay put, and the males have to wander around looking for a potential mate. The suggestion was that the darker males could get going faster (since we all know black objects heat up faster than light ones) and find the females first.

Sure enough, when Dr. Punzalan started heating bugs in the lab, the darker ones got going a little before the lighter ones, suggesting that they might have an edge. And then, in mating experiments, he was able to show that if the air temperature was warm, all the males did well, but if they simulated a cooler day, it was the darker-coloured males who could get up and find the females first.

When they find a female, Dr. Punzalan says, "Males do guard females. They hunker down and clasp them on the top. This prevents other males from getting access to these females. So what happens is that these darker males actually get that edge. Kind of like the early bird gets the worm." It seems that being dark and able to get hot really does work for these insects.

Which brings up the question: why isn't every male dark? Well, these solar panels, as Dr. Punzalan describes the insect's dark patches, are energetically expensive to produce. Only the healthiest, strongest males can get dark, while weaker males stay pale. Being dark is better, but it comes at a cost.

It seems that when it comes to hot sex, the ambush bug is one species that takes the expression literally.

QUIET CRICKETS

It's well known that male crickets serenade the females as part of their mating ritual. The females appear to like the rasping noise, and move in close to the male. Which is why Dr. Marlene Zuk, a behavioural ecologist at the University of California, Riverside, was so surprised when she realized that she couldn't hear any crickets on the Hawaiian island of Kauai. They seemed to be still alive and doing well, but something was shutting up the males.

It turned out that their sweet song had been attracting more than just females, and the males had learned to keep quiet to save their own lives. The culprit? A parasitic fly that's attuned to the sound of the males. It's a gruesome story. When this parasitoid (*Ormia ochracea*) hears a male field cricket (*Teleogryllus oceanicus*) singing, it homes in. The fly then deposits larvae on the cricket and, well, Dr. Zuk can tell the next part of the story. "The larvae then burrow into the cricket and grow inside of it while the male is still alive, so it's all very gruesome and alien-like. After about a week, the fly larvae burst out, and kill the cricket in the process." Let's move on, quickly.

Here's the problem for the male cricket. He needs to attract a mate, and in the field, the best way to do that is to call out so she knows where you are. But at the same time, you don't want to get infected by one of those larvae, so you're better off keeping quiet. Luckily, on Kauai, the males seem to have come up with a solution.

Discovering what it is was something of an accident. Dr. Zuk had been studying these crickets for a number of years, and in the late 1990s, she noticed the numbers of crickets was in decline. Then, in 2001, during her field study, she heard only one male calling (which she caught), and figured the crickets were on their way out. But then in 2003, "We went back just to check on this," she says, "and indeed, I heard nothing and thought, well, okay, but you might as well at least get out of the car. And lo and behold – in my headlamp of my car – I started seeing crickets. You have to understand that for someone who studies crickets, to see a bunch of them but to hear none is deeply disturbing."

Obviously the crickets had found a way around the problem. Their solution has two parts. First, there's the issue of why so many of the males are silent. This is a problem with a genetic solution. It turns out that between 90 and 95 per cent of the males on Kauai have a mutation that changes their wings, and crickets produce their sounds by rubbing their wings together. This mutation takes away the part of the wing used for making their song (Dr. Zuk calls these males "flat-wings"). The second part involves the remaining 5 per cent of these critters. The key is that these males do sing. And when they do, it attracts other males, as well as the females. This results in what Dr. Zuk describes as "a singles bar," with all the males and females flocking to the lone singer, and finding each other instead. Apparently, these few singing males are enough to keep the population thriving.

This situation is very specific to the island of Kauai. The same species of cricket is found on the other Hawaiian islands, but there are fewer of the flies, so the flat-wing crickets haven't evolved there. If Dr. Zuk plays the local cricket song on Kauai, then crickets flock in. But if she does the same thing anywhere else, then she's just shouting in the wind. As long as they have their own call, males aren't interested in hanging out where other guys can be found.

The other amazing part to this story is how quickly the mutation has spread through the population. Dr. Zuk estimates that it has taken less than twenty generations for this high level of change. What remains to be seen is what will happen over the next twenty generations. It's possible there will be some cycling: for a while, the chirping crickets will do well, as they attract more mates, which will increase their numbers in the population. Then the flies will become more populous, since they'll be able to find more of the male crickets. Which will lead to a rise in the number of flat-wings, and so on.

Or the males could give up singing altogether and just persuade Dr. Zuk to set up more chirping stations where they can gather. Now, there's a future career for someone: DJ-ing for insects.

Sniffing for Sexual Satisfaction

Most humans do not consider our body odour to be a pleasant smell, and we've come up with all kinds of potions and lotions in order to hide our scent. But for insects, their body's smell is often key to their communication. Their scent is made up of different compounds called pheromones, which are used in different combinations and at different times to signal all kinds of messages, from where to eat to when to mate. It isn't enough, though, just to produce a chemical message. If it's sex you're interested in, your potential partner has to be able to sense the aphrodisiac on the wind. So certain insects have developed unusual ways of picking out the perfume. Dr. Catherine Loudon, an evolutionary biologist at the University of California, Irvine, tied down some silkworm moths (*Bombyx mori*) to pin down their strategy.

If you're interested in seeing a silkworm moth, don't bother looking at any moths flying by you. These slightly chubby white moths, two and a half centimetres (one inch) long, don't fly, even though they have wings. These critters spend their lives just walking around. Nor do they eat. For them, finding a mate is their sole reason for existing. Oddly, though, the males still flap their wings vigorously, as if trying to fly, which was the first clue for Dr. Loudon about the moth's strategy for finding a mate.

The logic goes like this. An insect looking for a chemical cue is more than likely to use its antennae, which are very sensitive structures, covered in tiny sensory hairs. How tiny? According to Dr. Loudon, "The individual sensory hairs are about two microns in diameter, which is a fraction of a red blood cell in a mammal." These antennae are how an insect "smells" the world around it.

But if the air is still, then there's not going to be much scent for the insect to pick up. Think about your own nostrils. If you want to pick up a smell, you can either wait for it to drift by on the wind, or you can take a deep breath, pulling air up your nose, past your smell sensors. The male silkworm moth can't take a deep breath to drag air past its antennae, and waving the antennae around is only going to be of limited help. So how does he generate a breeze? He fans his wings, of course! This generates enough of a breeze that he can feel the air moving over his antennae. Now the males, who are the ones who seek out a sexual partner, are able to pick up the scent of any lovesick females nearby.

To test this hypothesis, Dr. Loudon had to tether male moths, "because," she says, "they tend to run around a little bit as they're doing this." After tying them down, she was able to use an instrument called a hotwire manometer to measure airflow, and show that these males really are generating enough of a breeze to stimulate their antennae, and that this self-generated wind is critical to their finding mates. When Dr. Loudon prevented the males from flapping, they couldn't find the females – even if they were right in front of them. They could see the females, but without the breeze across their antennae, they couldn't tell that they were potential mates, so they kept on moving. This might seem strange to us, but as Dr. Loudon puts it, "We're such highly visual creatures, it's hard to put it in the right analogy. It would be as if we could see an image, but we weren't really sure where it was."

So, this flapping is critical to mate detection. It's also important to note that this fanning is highly directional. Not only will a male know there's a female in the neighbourhood, but he'll also know she's right in front of him when the scent is strongest. Which is a pretty important fact to have when you're interested only in finding a mate before you die. The fanning is very effective, in case you were wondering. Dr. Loudon has calculated that it improves the chances of finding a mate by up to 800 times.

This behaviour is probably not unique to silkworm moths. Lots of other moths flap their wings while sitting still. It's the kind of discovery that might have practical implications for humans, too. Robots that "sniff" the air are becoming more common. Maybe borrowing the silkworm strategy, and equipping them with wings, would also make sense.

WETA WALKING

For most of us, stumbling around in the jungle in the middle of the night, only to discover a hideous, long-legged insect as large as our hand would be enough to trigger a full-scale shrieking phobic attack. That's the kind of reaction that separates entomologists from the rest of us. Entomologist Dr. Darryl Gwynne, from the University of Toronto, Mississauga, would consider such an encounter the highlight of the field season. And he did, when he went to the islands of New Zealand in search of the giant weta. On several nights he stumbled across this beastly bug and was able to answer a simple, yet important, question: why do the males have such long legs?

The weta, as a group of insects, tends to get entomologists all excited. They're found primarily in New Zealand, although some species are found in other parts of the southern hemisphere. And they've developed all kinds of curious adaptations. Some have become really big, and one species is so large that it's rumoured, quite fancifully, to be able to bring down sheep. Others have developed, as Dr. Gwynne says, "bizarre weaponry on the males. They're very flashy bugs." The exact form of the weaponry on the males varies from species to species. Notably, one has developed tiny tusks for fighting other males!

All the weta look like really big crickets, which isn't surprising as they are related to the cricket. And by big, we're talking huge. The record for Dr. Gwynne is one egg-laden female that weighed seventy grams (two and a half ounces). Considering that a typical cricket weighs just one or two grams, that's a big difference. Even the more moderate-sized giant wetas weigh in at between twenty and thirty grams (around one ounce) for females and ten grams (one third of an ounce) for males, so this is still an insect look-alike that dwarfs the crickets of North America. Like crickets, weta can jump, or they are equipped to do so, but they rarely do. Because they weigh as much as they do, there's not a lot of jumping in their joint. And the New Zealand species are all flightless.

Their big size has put the weta in danger. Dr. Gwynne calls them "walking meat pies, because for predators [they are easy to catch] – they don't fly, they don't jump very well – so they're highly threatened on the mainland." This has become a particular problem since the New Zealand islands were colonized by westerners, who brought with them dogs, and pigs, and rats.

Today, the weta has been mainly extinguished on the North Island and is very rare on the South Island, although they continue to live on some of the smaller islands the predators never reached.

From a scientific perspective, wetas make a good model for studying evolutionary biology. They've developed in relative isolation in the southern hemisphere, and there are many different species to study. Dr. Gwynne was interested primarily in the giant weta, especially in the marked differences in size between the sexes – the males are much smaller than the females, about half the size. What advantage could this give to the males?

The key to the answer comes from thinking about where these insects live. They're in the forest, but quite widely dispersed. That means they have to do a lot of walking to find a mate. And, in most insects species, it's the male that does the searching. From other work, Dr. Gwynne knew that if the insect has a long way to go, then being small and light is better. The lighter it is, the less energy it takes to walk. So, he captured some male giant wetas and stuck radio tags on their backs. Now he could track them, and see the purpose behind their nocturnal meanderings. And not only their nights, either. Dr. Gwynne also wanted to know, he says, "who they're spending the day with. Because they mate most of the day when they find a mating partner," making him the entomological equivalent of the paparazzi.

From following the insects around, he was able to confirm that the smallest males were the ones that had the most mating success. Small males also have relatively long legs for their size, so it seems that during their night wandering they are more likely than the bigger, squatter males to encounter a female, presumably because they can walk farther.

The next part of Dr. Gwynne's work is a little voyeuristic, be warned. But the way he confirmed mating success was by figuring out where the insects went after they met. A male weta, once he finds a female during the night, will follow her until they settle down for the day in a spot at the bottom of a plant. "They get very cozy together in a sort of a spooning situation," Dr. Gwynne says, "and they'll spend the whole day together in that little secluded spot. And they will mate frequently, up to fourteen or fifteen times a day." How does he know they mate that often? The male gives the female a gift, a little package of sperm, each time they mate, and yes, Dr. Gwynne sits there and counts them.

These males really do need to travel. Dr. Gwynne found that some of the smaller males, with the longer legs, were travelling as far as eighty metres (eighty-seven yards). That would be the equivalent of a human walking between six and seven kilometres (three and four miles) every night, which is a pretty hefty distance to travel on foot to find someone to sleep with. It would certainly be an interesting change to humanity if everyone had to wear out their shoes every time they wanted a date.

Cricket Coupling

Imagine you're a female cricket. It's Saturday night, and you and your cricket friends are drinking down at the local insect watering hole, the Stagnant Pond. A male cricket walks into the bar. He's buff and tanned, with three pairs of great legs. You think he's the kind of arthropod you could really go for, but then he gets closer and you sense that there's something off-putting about him. Then you realize it's because the two of you have already had a one-night stand, and suddenly you're no longer interested.

Okay, it's true that crickets don't hang out in bars, but the rest of this picture is accurate. Female tropical house crickets (*Gryllodes sigillatus*) are picky creatures and prefer not to mate with the same male twice. But how can a female tell whether she's been with a male before? That's the question Dr. Tracie Ivy wanted to answer. (She was a Ph.D. student at Illinois State University at the time, and is now a post-doctoral fellow in biology at the University of Rochester.)

It started with her research program, which involved watching crickets mating. "I've watched so many cricket matings that I don't even know if I could begin to add them up," she says. And during all these observations, she noticed two things. One was the female's reluctance to mate with the same male twice. The other was that those who were obliged, by lack of choice, to mate with the same male over again would have fewer offspring than those females who had a variety of possible mates. So, being choosy did have an advantage for the females. But how did they remember which males to avoid? As Dr. Ivy points out, "Their brains just aren't very big. I don't mean to shortchange them, but it seemed like there had to be a simpler mechanism."

We're talking about insects here, and key to arthropod communication is chemistry. Not the kind that happens between two infatuated humans, but the real, lab-based, test-tube kind. There are many ways that insects use chemical cues to get messages across, which set Dr. Ivy to thinking. There were two possible ways the crickets could be communicating. "Our first idea," she says, "was that perhaps each individual cricket has its own sort of chemical signature. What we thought might be happening is that females are leaving their own chemical signature upon the male with whom they've mated." The reverse could be true, too – that the male had his own signature, and that's what the females were picking up and remembering.

To test whether it was male or female crickets that were leaving their body odour behind required a complicated set-up.

First, Dr. Ivy needed to breed a line of male crickets that were genetically the same, and also a female line of basically cloned crickets. If the crickets had the same genes, they'd smell alike, too. After a while, Dr. Ivy had a group of males with one smell, and females with another. Now it was time to mix and match.

Experiment One went like this. A female mated with a male on the first day. On the second day, a new female was given the chance to mate with that male, or with another. Dr. Ivy watched to see which male she chose. This experiment tested whether the female was detecting her own scent on the male. Remember, even though this was a new female, she had the same scent as her sister, so if she was detecting that scent, she'd pick the stranger male, not the one her sister had mated with the day before.

Experiment Two tested the opposite. A female was mated with a male on day one. On the second day, the same female was introduced to a brother of the male she'd mated with the day before, and also to a male that was unrelated to the male from day one. Dr. Ivy's idea was that if the female avoided the sibling of the mate from the day before, then it must be that she was recognizing his smell.

As you can imagine, this experiment was a nightmare to run several dozen times. "I tell you," says Dr. Ivy, "it was pretty difficult to keep all these guys straight." But at the end of the study, she had her answer. "We found that the female is marking the male, and we found no support for the other idea, that females are actually remembering and recognizing the males with whom they've mated. So the female is sensing her own scent on this male that she has previously mated with and kind of saying, 'Oh, I've been there. Don't want to go there again.'"

Exactly what the chemical signal that the female is picking up on isn't known at this point. Since she mounts the male during mating, though, it's probably something she's leaving behind after their sexual connection – that would be the simplest answer. But whatever it is, the fact that the female recognizes her own scent is a satisfying answer for Dr. Ivy. "At first," she says, "I thought that females were probably recognizing the males. But after we got our results and I had thought about it more, I thought, well, you know, it's a lot easier for a female to remember herself because it requires a lot less brain power on her part. I mean, she always has herself there to make a comparison."

Same as us, really. We might not consciously recognize our own body odour, but we do know it, even if we spend a lot of time washing it away or covering it with perfumes. Happily for married couples, we don't have the same predilection for novelty that the crickets do. Otherwise, men would be spending a lot more time in the shower, washing off their partner's smell.

Brotherly Love Among Crayfish

Ask people what animal is mostly closely linked with peace, and most of them will answer "the dove." But if you're looking for something that epitomizes peaceful resolution, then you might want to pick a different creature. While it might not have the beauty and poetry of the dove, the red swamp crayfish (*Procambarus clarkii*) is a better choice for showing the meaning of "love thy neighbour." According to Dr. Donald Edwards, a neurobiologist at Georgia State University, the males of this species, when they want a permanent end to conflict, spread the love.

If you've ever chowed down a plate of Louisiana crawfish, then you know what these little crustaceans look like. If not, then imagine a bite-sized lobster. They live in swamps, mud puddles, and river bottoms throughout the southern United States, and also on dinner plates wherever Cajun cooking is served.

Life for these critters isn't easy. In their muddy environments, there's only so much food to go around, which can lead to fights. Like many other animals, these crayfish live in hierarchies, with the most dominant crayfish getting the best food and the sweetest living quarters. The most common way for hierarchical animals to figure out who's at the top of the heap is to fight. And these beasts will do that. Two males will start to tussle, and then get into a full-on scrap. Eventually, one will back off and try to escape, which, when they're in Dr. Edwards's lab, isn't easy. The other male will harass him for a while, but in time they'll settle down. And from then on, the male that backed off will be subordinate to the other animal.

Except, that isn't always what happens. Dr. Edwards had a graduate student working with him in the lab, who saw something odd. "He noticed one day," says Dr. Edwards, "that several pairs seemed to be engaging in sex, seemed to be copulating – and that was odd because all of them were males."

Now, crayfish sex is pretty obvious when you see it. Let Dr. Edwards describe what happens: "It's remarkably like people. It's face to face. The male will approach the female from the side or from behind and clasp her with his walking legs. She will relax and turn over and put all her legs and her claws forward, and they face each other. And he will grab her claws with his claws to hold them up and out of the way and hold her with his legs. Then he extends his abdomen along hers. He's got an organa, a cuticular organa, which looks rather like a fountain pen, through which he extrudes sperm, and he'll put it right over where she's got an ova duct."

Except, in this case, there weren't any females in the tank. Males were going through exactly the same motions with a second male, including depositing sperm on the other male. It didn't last as long as male-to-female sex, however. This pseudo-copulation, as it's called, lasted for a few minutes, while real copulation can last for the better part of an hour. But it did seem to have an effect on fighting. If a subordinate male engaged in pseudo-copulation with a more dominant animal, then the aggression among all the crayfish in the tank would drop off quickly, within about fifteen minutes. On the other hand, if the subordinate chose to fight instead, then the dominant crayfish would batter him – for up to an hour. As Dr. Edwards puts it, "What pseudo-copulation seems to do is to signify for both animals that they've reached an agreement as to which animal is dominant and which animal is subordinate. It's the future dominant that plays the male role in this interaction and the future subordinate that plays the female role."

There's a dramatic consequence of this behaviour for the subordinate, too. In half the animals Dr. Edwards studied, if the subordinate resisted and fought back, or never got the chance to participate in this sexual behaviour, then within a day he would be found dead. For crayfish, there seems to be a stark choice between submission and termination – which seems tough, but presumably helps keep the peace.

It's not that surprising a behaviour to Dr. Edwards. In the insect world particularly, sex and violence are often related. Among spiders, for example, the male will often bring a gift when he comes to mate with a female. It's a way of saying, "I'm not here to hunt you, I'm only here for sex." Without the gift, one of them is going to get chomped on. The same kind of relationship exists among other invertebrates. Many of them are more likely to fight another of their own kind when they encounter one than to make friends. If the species is going to survive, that fighting instinct has to be turned off for encounters with a member of the opposite sex. The crayfish seem to have made this a more general behaviour. So any crayfish that flips over in submission is a potential sex partner, and not a rival you're going to try to kill.

Crayfish give the slogan "Make love, not war" a whole new twist.

Bisexual Beetles

Among humans there are those who view homosexuality as primarily either a moral or a political issue, and there are some who even argue that it goes against nature. But ask any biologist and she'll tell you that homosexuality is not only perfectly natural, it's quite common in the animal kingdom. As common as it is, though, it does leave scientists scratching their heads. After all, if the point of sex is to pass genes on to the next generation, exactly what purpose does homosexuality serve? Well, Dr. Sara Lewis, an evolutionary ecologist at Tufts University in Massachusetts, thinks she's found at least one answer. In male red flour beetles (*Tribolium castaneum*), homosexuality may be just a roundabout way of passing on their genes anyway.

Red flour beetles are relatively common throughout the world. They live, as their name suggests, on flour products. They can infest all kinds of grains and nuts, although you'd likely not notice them if they were in your kitchen. The adults are only a few millimetres long (roughly one seventh of an inch), about the size of a small grain of rice. They're brown and typically beetle-like in their appearance.

Because they're so small, they're not the easiest beetles to study, particularly if you're trying to observe their sexual activity.

But, with true voyeuristic resolve, Dr. Lewis put pairs of male beetles in tubes under a microscope, and sure enough, they performed as expected. What was a surprise was to find that the males were not only going through the motions of sex, they were actually releasing sperm.

The big question is why spend energy on a sexual activity that's not going to lead to any offspring? Various theories have been put forward, but not many of them have been scientifically tested. One suggestion is that males can't tell the difference between other males and females. Or, as Dr. Lewis put it, "It's a case of mistaken sexual identity." Another suggestion is that it's all about dominance: sex as a weapon to keep other males in their place. Then there's the idea that rather than dominating other males, it may be a way to increase group cohesion, bringing the males closer together in more than just a physical sense. Or, it could be that there's an advantage to being hypersexual, and homosexuality is just one aspect of increased sexual activity. Or finally, it could be that the males are practising, so that they're ready when a female comes along.

Dr. Lewis decided to try to sort out which among all these hypotheses fit the red flour beetles. Dominance? She found no evidence of that going on among these beetles. And they don't have much of a social structure, so group cohesion didn't make sense either. The "practice makes perfect" idea didn't hold water either; the beetles weren't any more successful with females if they'd had previous homosexual encounters. But there was one other hypothesis that Dr. Lewis wanted to try out. "We decided," says Dr. Lewis, "to test the idea that males might somehow be gaining a subtle reproductive benefit by indirectly transferring sperm during these homosexual copulations." Remember, Dr. Lewis had already discovered that when these males were having sex with other males, they deposited sperm as part of the activity. Could it be that this sperm was somehow making it to females anyway? This would be the indirect transfer she refers to.

To test this, she observed males mating and then separated them, allowing both of them to go on and mate with different females. The offspring were allowed to hatch, and then it was time for paternity tests. In 7 per cent of the females, some of the offspring were the progeny of males she'd never had contact with. The evidence suggested that sperm was transferring from one male to another, and ultimately to the female.

This wasn't the result she was expecting. "We were really surprised by this. In fact, it was such a surprise that we kept running this particular experiment over and over again. We spent about two years repeating this experiment just to make sure that this was really happening."

How, exactly, this happens remains a mystery. The one male deposits his sperm on the outside of the second male, not in the reproductive system, so there's no obvious way for it to get into the female. But the numbers are too high for this to be just chance. And it does explain, for these beetles anyway, why homosexuality might persist in the population. "This is kind of a hedging strategy," says Dr. Lewis. "If males never have a chance to encounter a female, then they can still ensure that some of their genes get into the next generation by mating with other males, who then indirectly translocate their sperm to the females that they subsequently encounter. So a male can sire progeny with a female that he's never even mated with." As for siring offspring with females he mates with directly, well, that's just icing on the cake.

It still remains to be seen if this explains homosexuality in any other creatures, but for the red flour beetle, it seems that there's a real evolutionary advantage to the old line, "Love the one you're with."

Cockroach Wars

If you were using an insect dating service, here's the kind of ad you might place: "Single, young, female cockroach seeks male cockroach for a good time. Must be underdeveloped, slightly thin, and not overbearing. Poor genes are a definite asset." That's what Dr. Allen Moore, an evolutionary geneticist at the University of Exeter in England, believes you'd write. Unlike most other creatures, the female Tanzanian cockroach (*Nauphoeta cinerea*) seems to prefer wimpy males over the big, strong types.

You might be wondering how to tell a wimpy cockroach from a brawny one. It turns out it's not that hard. In a crowd of cockroaches, there will always be bullies. Like in a schoolyard, they're pushing their weight around, bossing the other cockroaches, biting them. If the other cockroaches had lunch money, the bullies would probably steal it. And the wimps? They're the ones running away and hiding.

Cockroaches aren't particularly visual critters, so, for the females, telling the wimps from the bullies is a chemical matter. Males produce a pheromone made up of three different chemicals that the females can sense. The balance of these three chemicals is what the female is sniffing for, because the wimpy males produce far more of one them than the dominant males do. For the females, that's the aphrodisiac that'll make her pick her man.

Here's an interesting twist, though. You'd think the dominant males would come up with a way to fake out the females – perhaps produce more of this third chemical, to make themselves smell wimpy and get her attention. But when Dr. Moore tried to fool females in this way, he found something he hadn't expected. "It probably isn't a system that can be cheated," he says, "because when we apply the compound to the males, we actually cause them to become wimp-like. They can't just smell that way. They're going to act that way as well. Their odour actually causes their behaviour to change as well." A case of "you are what you wear."

The obvious next question is why females prefer the wimpier males. In most species, strength is what gets the female's attention. The answer isn't known at this point. But what is clear is that females who mate with wimpy males produce fewer sons. This could be an advantage to the sons, since, if there are fewer males around, they're going to be more attractive to eligible females. It might also be safer for the females to mate with less aggressive males, for their own health, anyway.

If the females always got their way, it would eventually lead to a population made up entirely of wimpy males (the wimpy/aggressive traits are inherited from the parents). But that's not what happens. The bullies do get their way as well. The females might be searching for wimps, but the aggressive males are there to push their puny rivals away and get the female for themselves. As Dr. Moore says, "You can't predict from one generation to the next who's going to win, so it cycles. Sometimes the females win, sometimes the males win, but it's a tug-of-war, really. It's called a sexual conflict, where sometimes what's best is to be the most attractive male, and sometimes what's best is to be the biggest, bulliest male."

This sexual conflict is a bit of a confusion for evolution. The females are acting in their own best interests by trying to find the softer males. The males, on the other hand, are more likely to land a partner when they're pushy. Overall, then, this means the species will produce fewer offspring than it would if males and females had the same goals. But it's pretty common for there to be conflict between the sexes, so it must have some value.

And while researchers try to figure that out, guys who are less aggressive, take heart. Females find vulnerability sexy. At least, female cockroaches do.

Water from Weevils' Sperm

Most people find insects somewhat unpleasant, but maybe they'd change their minds if they knew that insects are actually a pretty romantic lot. After all, the males frequently give their potential mate a gift: food is pretty common, though sometimes it's just flowers, or even a stone. But these gifts are more than just an expression of affection. They may be a way of securing fidelity from a mate with a wandering eye – or antennae. That's the case for the cowpea weevil (*Callosobruchus maculatus*). Dr. Martin Edvardsson, now at the Australian National University in Canberra, has discovered that this tiny insect keeps his mate from straying with a unique gift.

The male achieves his goal by first producing a shocking amount of seminal fluid. By shocking, we're talking about 10 per cent of his own body mass in a single ejaculation. "In human terms," says Dr. Edvardsson, "that would be about seven or eight kilograms [sixteen to seventeen pounds] for the average-sized man." Which, you have to agree, is a lot. Remember, though, this is an insect, so it has a hard exoskeleton, meaning that losing all that weight at once doesn't cause the weevil to shrink. But because they are quite a bit lighter after sex, Dr. Edvardsson was able to figure out the ejaculate size. He would weigh the male before sex, and then again afterwards, and calculate how much weight he'd lost. (Dr. Edvardsson made his discoveries while working with Allen Moore, the scientist at the University of Exeter who just introduced us to the cockroach wars.)

To understand why the male would expend so much liquid on

a female, we need to talk about the insect itself. These are fairly typical-looking little beetles, just a few millimetres (a fraction of an inch) long, and coloured brown and black. They live on beans – the cowpea and black-eyed pea are among their favourites – and they're generally found in warm, dry places.

It was their habitat that started Dr. Edvardsson thinking about why males would give up so much liquid during sex. His hypothesis was that this gift would make the male more attractive to the female. And that might have to do with how thirsty the female was. So, he set up an experiment where some of the females could get as much water as they wanted from the environment, and placed others in a much drier place.

In this species, the female has a wide choice of mates. When a male approaches a female, he taps her on the back with his antennae. If she agrees, the male mounts her from behind. Sex takes place, with the male attached by his genitalia, and he deposits his massive ejaculate.

In the experiment, Dr. Edvardsson found that the females with access to lots of water were far less interested in mating than the females in the dry environment. What he believes is that "they derive water from the ejaculates, because they live in really arid areas, and water is probably a really important resource to them. And they can somehow then get the water from the seminal fluid and use that." While it's not known exactly how much of the seminal fluid is water, if these insects are like most species, it probably makes up the bulk of the fluid. And the mechanism for the female to absorb the water is also still unknown, but the data show that they are absorbing it somehow.

What does this have to do with preventing promiscuity by the females? It may be why the males produce such a giant ejaculate. After all, unlike most gifts, the female can't see beforehand what the male is going to give her. And if she's given a big water gift, then she's not going to need to drink for a long time. If that's the case, she won't go looking for another male to mate with, because she'll be sated. From the male's perspective, this means his sperm will have more time to fertilize her eggs, and he'll father more of her offspring.

There's more to this story than just the water gift for the female. The male's penis is covered in spikes, and mating causes damage to the female's genital tract. So she's not going to mate until she really needs water, again giving the first male's sperm more time to fertilize her eggs.

So the male gets a faithful mate, and because the eggs are in a female who's got plenty of water, there's more chance they'll be healthy when they hatch. It makes giving up such a lot of liquid seem like a reasonable act, even though it's going to diminish his opportunities to mate. Remember, he's in a dry world too, so it's going to take him a long time to restore his liquid after he ejaculates.

Nothing like a long, tall drink of water to satisfy your partner.

LYING FLIES

If you're a young woman looking for a date, here are some pickup lines you might want to try the next time you're in a bar: "Oooh... do you work out?" or "Haven't I seen you here before?" How about, "Kiss me, I'm pregnant!" Maybe that last line isn't what most humans would use, but according to Dr. Natasha LeBas, an evolutionary biologist at the University of Western Australia, if you're a female dancer fly (*Rhamphomyia sulcata*), looking pregnant is just the message you want to send to a male fly to get his attention. This has led the females to evolve a complex ploy so that they look like they're carrying eggs, even when they're not.

Examine a female dancer fly yourself and you can see this ploy in action. Don't know what a dancer fly looks like? They're pretty common flies, found all over the world, and the males are nothing special to look at. They look like any other common fly, with slightly longer legs. But the females are the special ones. They hang out in swarms, dancing together, which is where their name comes from, and if you look closely, you can see that they look as if they have balloons hanging off their abdomens. Zoom in a little closer and you'll notice that their legs are covered with scales that look, as Dr. LeBas describes them, "frilly."

Now, not that many of us go around looking closely at the bellies of flies. But if you're a male dancer fly, the belly of a female is exactly what you're interested in. And these frilly legs, along with balloons hanging off the belly, make the female look full of eggs that are just ready to be laid.

For male dancer flies, there's no aphrodisiac better than a pregnant female. Dr. LeBas says, "Males want females that have lots of mature eggs but the eggs aren't fertilized. So they want the females to be as close to laying the eggs as possible, the eggs to be mature, and for the female to have lots of them. So then the male can fertilize these eggs, and he'll get to be the father." This might seem odd, but in this species, the female can mate whenever she wants but doesn't fertilize her eggs until just before they're laid. That means, from a male perspective, you want to be the last male she meets, since your sperm will be the ones she's most likely to use.

Why would a female want to mate more than once, if she's only going to use one set of sperm to fertilize her eggs? It turns out it's all about food. Male dancer flies, like many insects, don't arrive empty-handed when they come to mate with a female. They bring a present, called, appropriately enough, a nuptial gift, in most cases food. The successful male gives his gift to the female – say, a nice, juicy, nutritious insect – and then while she eats, he mates. In a situation like this, everyone wins. She gets food to have the energy to make more eggs; he gets to be a dad.

Remember, though, males of this species are reluctant to mate with females until they're close to laying their eggs. This creates a problem for the females. The best time for them to get food is when they're just producing eggs; she doesn't need as much energy later on. Hence the lie. Persuade the male to mate with you when it's more to your advantage than his – that is, before the eggs are mature – and, as a female, you're going to get what you need. Some species have taken this to an extreme: the females have lost the ability to hunt and are totally dependent on males for all their food! Adds a whole new dimension to the idea of bringing chocolates and flowers on a date.

This lie by the females isn't that serious, though. In a surprise twist, the females that look the most pregnant, with nice frilly legs and big abdominal balloons, are also the most fertile. So the message they're sending to the males is not wholly inaccurate. They might not be pregnant right now, but when they fill up with eggs, there will be plenty of them, which means those males are hitting the genetic jackpot, with lots of possible offspring. It's a white lie, really, since everyone's going to do well out of this arrangement.

Not that males always play by the rules. Sometimes males cheat and only give stones or leaves or twigs. But, if females are dependent on the food gifts, that strategy isn't going to last long.

So, the next time you're in a bar, and you see girls pretending they're pregnant, and boys wandering around with boxes of chocolates, you'll know they're just imitating the mating strategy of the dancer fly.

Even Animals, Land and Sea, Do It

❖

From the plains of Africa to the deepest ocean, it's hard to find a species that isn't at least a bit picky about choosing a date. And what really turns on a possible partner can be something of a surprise . . .

Primate Copulation Call

Ever been in an apartment or hotel room with really thin walls, and really loud neighbours? Ever been woken up in the middle of the night by their lovemaking? Then this story will seem familiar. If there's one thing a female yellow baboon (*Papio cynocephalus*) isn't afraid of, it's expressing her sexual satisfaction. You might even say she advertises it. For her, the better the boy, the bigger the bellow. Dr. Stuart Semple, a primatologist at Roehampton University in London, England, thinks he knows why these animals have such a raucous rating system. It seems it's a way for females to get the best mating possible.

When female yellow baboons are at their most fertile, a period known as estrus, they like to mate frequently. We're talking upwards of twice an hour, all day long. Each mating lasts only between five and ten seconds. As Dr. Semple puts it, "They're quick, but they do it a lot." It's what comes at the end of the mating, and shortly thereafter, that's really interesting.

Just as the sex is wrapping up, the female will start to grunt. These grunts are really loud. Dr. Semple has heard them from over 100 metres (110 yards) away. Not only does the female grunt, but she also runs away from the male, and keeps grunting for about fifteen seconds, longer than copulation took in the first place.

These grunts are very personal to each female. To the trained ear, it's possible to tell which female is calling. Male baboons, naturally, are attuned to this. And in experiments, Dr. Semple was able to show that the males know exactly which female is having sex. But the female is saying more than just, "Hey guys, I'm here, and I'm in the mood." She's also telling them how fertile she is. Baboons, like all apes, are more fertile at some points during their cycle than others, and through these grunts, the female is letting the males know what their chances are of being the one to get her pregnant. After all, as Dr. Semple points out, "There's no point necessarily mating with a female if she's definitely not going to ovulate. That's a waste of energy. He could be putting his energy into looking around for other females, or eating."

That's not the only information the female's giving out, though. Along with telling males that she's ready to breed, she's letting everyone know just how good her last lover was.

Baboons live in a very hierarchical structure. The top males can get any females they want. Males farther down the ladder have to work hard to get access to females – it's a fight if you want to breed. And when a female has just had sex with a big, powerful male, she puts much more energy into her call than she will with a puny, lower-ranking male.

So why bother with all this advertising? Well, female baboons aren't fertile for that long each estrus, so it makes sense for her to try and get the best bang for her buck (so to speak). She wants her offspring to have the best genes possible. This shout-out might help that.

Her bellowing at the end of copulation excites the other males around. They know there's a receptive female around, and what baboon wouldn't jump at the chance to mate? That's going to lead to scuffles between the males, as they all vie for her attention, with the strongest winning and approaching the female to mate. But why shout louder when the male you just mated with is strong? Dr. Semple thinks it's something of a challenge. "She, if you like, excites the other males more, and she's going to need to do that because the male she's with is very strong, and if the other males are going to depose him, they're going to need to compete very strongly." It's as if she's trying to goad all the boys into ensuring that she gets to mate with the fittest male possible.

A curious sidebar to this story is that females will, from time to time, fake this call. Sometimes a female is under attack from a male, but there's no sex going on. She'll run away then and make the sex call, which Dr. Semple thinks is her way of attracting the attention of other males in the area, who might protect her from the first male's aggression. Not that it happens often, but it does suggest dishonesty on the plains.

For yellow baboons, when it comes to sex, it pays to advertise.

SALACIOUS SIMIANS

It's been a long day, you get home late, no one else is awake, so you switch on the TV. Maybe you decide to watch one of those adult channels. Your excuse? Well, it's what a monkey would do. No, really, it's true. According to the work of Dr. Michael Platt, a neurobiologist at Duke University in Durham, North Carolina, macaque monkeys (*Macaca mulatta*) will willingly pay for the right to look at what amounts to primate pornography. Which doesn't just tell us about the sexual proclivity of monkeys, but may also explain some human behaviour.

By "paying" we don't mean exchanging cash. For monkeys, the currency is fruit juice. Macaques will do just about anything for fruit juice. They're so fond of juice that, given a choice between two different tasks, both of which will give them a squirt of juice, they invariably pick the one that's going to give them more juice, even if it's a job they don't particularly like. That is, until it comes to this primate porn. In those cases, the monkeys are willing to forgo juice altogether if it means they're going to get to see an image they like. Which Dr. Platt likens to them paying for the privilege.

What are these images? The short answer is they are pictures of the hindquarters of females they know, and the faces of high-ranking male monkeys. Dr. Platt worked this out using a fairly straightforward set of trials. Monkeys would be given a choice of either looking at a picture or getting a squirt of fruit juice. Some of the pictures were of male faces, some were female faces, and others were female rear ends. By changing the pictures, and changing the amount of fruit juice, Dr. Platt was able to compare which pictures were more significant than fruit juice, and which were the ones the monkeys weren't much interested in seeing. That's how he knows they're willing to give up, or pay with, fruit juice for the chance to look at a female derrière or a high-ranking male's mug shot.

You can probably figure out why the male monkeys wanted to look at the picture of a female's behind, but the motivation for looking at faces of high-ranking males is less obvious. It's known that high-ranking males hold a lot of influence in monkey society. They get first access to females and to the best food, and they tend to win in fights with lower-ranking males. But low-ranking males can get support and help from the higher-ranked animals. "So," says Dr. Platt, "you might imagine that it's very important for any monkey to keep track of where the higher-ranking animals are. You can imagine, as well, that low-ranking animals just don't have much influence, so it's not worth the time or the fruit juice to look at them."

It's the monkey equivalent of reading *Playboy* for the images of sex, and then reading *Forbes* magazine to keep up with the power-mongers. But there's more to this research. Dr. Platt says, "The really interesting aspect of our study is that, although these monkeys were willing to pay fruit juice to see images of high-ranking animals, when you look at how much time they actually spent looking at those images when they were displayed, they didn't discriminate between high- and low-ranking animals. They just looked away. However, when the female hindquarters came up, they looked at them for as long as they were displayed." So there's something different going on in the minds of these monkeys when they look at the sexual imagery compared to the pictures of high-powered animals. What that is isn't yet known, so that's where the research will head from here. Dr. Platt's plan is to look inside the monkey's brain to figure out what parts light up when the different types of images are shown. In the long run, then, the goal is to understand human emotion and motivation. Perhaps this will also explain why so many men say they're reading *Playboy* just for the articles.

BAT BRAINS AND BALLS

It's all about balance. We commonly think that if someone is gifted in one area, then she's probably lacking somewhere else. There's a stereotype that smart people tend to be clumsy and weedy, while the beautiful are not always so smart. These generalizations often don't hold, of course, but they're still out there. In the animal world, however, sometimes these trade-offs tell us important stories about evolution. Animals with any kind of exceptional feature usually need it for some important reason, but at the same time, that feature will cost them in some other way. Think of the giraffe. It has a long neck so it can reach up into the trees to get the best leaves, but that makes it harder for the beast to drink. Examples like this abound in nature, and Dr. Scott Pitnick, a biologist from Syracuse University, has found a particularly curious one, involving the brains and reproductive organs of bats.

Bats make a good group of mammals to study if you're interested in such trade-offs. As Dr. Pitnick explains, "They live their lives on the knife's edge, energetically, because they're small mammals and flight is very costly." This means any investment in brain size, or reproductive organs, or any other system, is going to have to be balanced by directing less energy elsewhere. They basically have nothing to waste.

Another reason for looking at bats is that they live in a variety of social structures. Some species live in groups of one male and several females. Other species live in groups where there are plenty of both sexes. Some species pair up, one male with one female. Not only are there all kinds of variations among the forms of groups, but there are different levels of promiscuity in the groups, too. Sometimes bat couples are monogamous. Sometimes the males mate with several females in a group, but the females mate only with that one male. In other species, everyone is promiscuous. This gives biologists plenty of different combinations to examine.

By looking at these different social and breeding schemes, Dr. Pitnick was able to find certain patterns. We'll talk about brains first. The biggest brains, relative to body size, are found in the species where both males and females are monogamous. Not far behind are the species where the females have one male to mate with, and the male has a limited number of female partners. The smallest brains were in bats that were most promiscuous. This wasn't what Dr. Pitnick had expected. "If anything," he says, "we were expecting the exact opposite relationship. The reason is that one of the prevailing hypotheses for brain size evolution is referred to as the social brain hypothesis. The idea here is that as social complexity increases, so does selection on individuals to subvert the needs of others in favour of your own selfish reproductive interests or just general interests."

In other words, the bats that are promiscuous should need bigger brains because they have to work harder to convince other bats to mate with them. Monogamous bats don't need to work as hard to keep their mate, so, in theory, they could get by with less brain power. But that's exactly the opposite of what Dr. Pitnick observed.

That's when he turned to looking at male reproductive organs, their testes. "When we started looking at the same kinds of breeding-system categorizations and how that related to testes mass," Dr. Pitnick says, "the graphs produced essentially the mirror opposite to that of brains. What we found are significant trade-offs that, evolutionarily, whenever you get a transition to female promiscuity, what we see is a decrease in relative brain mass associated with an increase in relative testes mass."

Or, put simply, the bigger the balls, the smaller the brain. Here's the trade-off at play. There's only so much energy to go around for building body structures. While most biologists have assumed that bigger brains are better, sometimes there might be something else that's more important for survival, and that's going to limit brain size. In this case, it's the testes.

Why would big balls be important to bats? It all comes down to sperm competition. When females mate with more than one male, then there's not only competition to be the male who gets access to the female, but the contest continues inside her body. Now it's the sperm's turn. It's to a male's advantage to have either the fastest, the strongest, or just the most sperm. That way his will be the ones that get to the egg and win the reproductive race. The way to do this is to have big testes that can produce lots of sperm. Hence, the bigger testes of males in the promiscuous species.

In the monogamous species, there's no competition, so there's no need for big balls. Instead, these species use their energy to spare to build bigger brains. After all, there are good reasons for having a big brain. It helps you find food faster, if nothing else.

While Dr. Pitnick was surprised by his results, that wasn't the universal response. "Only males seem to be surprised." And it does add some merit to the insult, "He thinks with his gonads."

AMOROUS ANTELOPE ANTICS

If there's one observation you could make about the Topi antelopes (*Damaliscus korrigum*) of Africa, it's that they never had to have a sexual revolution. Humans think of the sexual revolution as the time when women cast off their traditional roles and behaviours and could become the initiators and aggressors in the great human mating dance. A woman could choose a man and make it clear to him that he'd been chosen. For the Topi, though, that's all old hat. As Dr. Jakob Bro-Jørgensen, now a wildlife biologist at the University of Liverpool, has discovered, the Topi antelope is yet another example of a species that got there before us.

Topi antelope are found on the open plains of Africa. They're a medium-sized antelope, a relative of the wildebeest, mostly tan-brown in colour with darker patches on their legs and heads. Unlike other antelope species, the males and females both have long horns and look more or less the same. As Dr. Bro-Jørgensen says, "There's a very slight difference, but to the uninitiated, they look similar. You wouldn't be able to tell them apart."

What's particularly interesting for biologists about the Topi is their mating system. The males gather together in what's called a lek. The term "lek" comes from Swedish word for "play" but has been turned by scientists into the name for a gathering place where animals come to mate. Some animals are very careful about their leks, but there's nothing special about the places chosen by the Topi. It's not as though they're feeding spots or water holes – they're just open spaces on the savannah. But groups of males gather there, and each will defend a small territory within the lek. When the females are ready to mate, when they are in estrus, they come to the lek and try to mate with the males at its centre.

Since the females prefer the males at the middle of the lek, this is obviously the premium spot. So the males fight to win the centre spot. This fighting can be quite vicious. Dr. Bro-Jørgensen and his team have noticed that the victorious males tend to be the largest, but they're often covered with bleeding wounds from the fights for supremacy. And even once they win, the dominant males' problems aren't over.

Now the females get into the action, and they're the ones picking which male they'll mate with. The males are stuck defending their particular territories, so if a female isn't interested, all she has to do is leave and find another male. But here comes the problem. The female is interested in mating as many times as possible with the best males she can. The male, on the other hand, would like to, well, spread his seed around and mate with as many different females as he can. So there's a conflict here.

It's an interesting spectacle to watch. "The females are highly promiscuous," says Dr. Bro-Jørgensen, "so they mate not only with several males, but they mate repeatedly with each one. That means that there's quite a high demand on the male sperm supplies in the centre of the lek." As for the male, he says, "He usually goes for the female with whom he has mated the least so far. So he prefers novel females." And now the aggression begins.

The females will actually attack males who are mating with other females. "They run at the male," says Dr. Bro-Jørgensen. "They lower the horns, and sometimes they prick his bottom with the horn tips. Then what happens is the male jumps out of the way, and it often disrupts the mating." Which is understandable. Not many males can keep mating while jumping away because their butt is being prodded. And sometimes the male is annoyed enough that he'll attack the female who's poking him. But more often the first mating will end, and the male will turn his attention to the female who prodded him.

That's not the only fighting going on. The females fight one another as well, with the dominant animal usually winning and so getting more chances to mate. Not that the male cares whether the female he's mating with is dominant or not. He just knows how many times he's mated before with that female. Nonetheless, harassment works, and the deciding factor for which one he mates with the most is who's the pushiest.

It's still a battle of the sexes, but one that's turned the old idea on its head. When it comes to life on the African plains, it's the women who win.

Sea Slug Orgy

Looking for great sex? How about a wonder drug that's part Chanel No. 5 and part Viagra? One that will attract mates from miles around and boost their sexual performance? It sounds good, but this particular aphrodisiac does have one small hitch. The love potion only works for members of the genus *Aplysia*. For those not in the know, these creatures are commonly called sea slugs, but fans of this particular group of sea slugs call them sea hares.

Dr. Gregg Nagle, a cell biologist from the University of Texas Medical Branch in Galveston, is a fan of sea hares. He's been studying the love drug or pheromone they produce and how it helps them find romance under the seas.

Aplysia are found almost everywhere around the world, excluding the really cold waters of the Arctic and Antarctic. They're pretty big as sea slugs go, measuring anywhere from ten to twenty centimetres long (four to eight inches) and weighing as much as one kilogram (two and a quarter pounds). The hare part of the name? That comes from the tiny tentacles growing at the front of its body, which look like a pair of rabbit ears (if you have a vivid imagination).

Most of the time, these are solitary creatures, living out their lives on the sea floor. They live for only a year, so finding a mate during that time is very important. And that's tricky for the sea hare. Even though they're quite large, they live far apart, and they have really lousy vision. We can look around and pick out a potential partner, but that's just not an option for these animals. So instead they use pheromones.

Pheromones are chemical signals produced by animals, and they are often associated with sex. This pheromone's name gives away its function. It's called attractin, and its job is (obviously) to attract a mate. This is one potent molecule, somewhere between a hundred and a thousand times more potent than any hormone that humans can produce. Or, as Dr. Nagle says, "Just to give you an idea, you could take a teaspoon of this material and put it in a swimming pool, and that would be enough to attract all of the *Aplysia* in the neighbourhood."

Once the love drug has brought all the local sea hares into the neighbourhood, then the action gets going. And boy, does it get going. According to Dr. Nagle, "Between two animals, this mating can go on for several hours. But if you have an aggregation of five, fifteen, twenty animals, this aggregation can go on for days, so it's a long-term party. Once the party gets started, it just goes on and on."

Stamina isn't the only thing that keeps the party going. Sea hares are hermaphrodites. When they join the party, incoming animals behave as males and supply the sperm. Once the sperm starts to run out, then they'll switch to laying eggs, leaving other hares to provide the sperm. You'll end up with these giant gatherings, with sea hares flipping back and forth between mating as males and then laying eggs as females, all the while releasing more pheromones to get others to join the group action. It's a giant undersea orgy that *Aplysia* have been performing for hundreds of millions of years.

You might wonder why, if these animals are hermaphrodites, they don't just fertilize their own eggs, rather than going to all the trouble of attracting others. Dr. Nagle thinks they have to find others because they're incapable of fertilizing their own eggs. And if you're going to send out a message anyway, might as well make it potent.

Of course, a message like this could attract the wrong types as well. And, in this case, at least in the lab, it does seem that it gets the attention of a number of different *Aplysia* species. So, how do these myopic creatures tell one another apart? Dr. Nagle isn't sure, but the key is probably that these species wouldn't ever encounter each other in the wild, so it's unlikely to be a problem. It's possible that attractin also calls in predators, but sea slugs have an answer for that. They taste terrible! Researchers have tried cooking them, but they have bromides in their skin, which make them a rotten choice for a meal. Plus, they release big clouds of purple slime when they're attacked.

Aplysia: great lovers, but lousy for lunch!

THE BARNACLE'S PENIS

You'd think the lowly barnacle (*Balanus glandula*) wouldn't have much to brag about. After all, barnacles are the homebodies of the intertidal zone, confined to their humble shells and anchored for life to whatever rock they've settled on. But they are endowed with one trait that puts all other sea life to shame. Chris Neufeld, a Ph.D. candidate at the University of Alberta, has been studying the long and the short of the barnacle's reproductive system and found that these crafty crustaceans are actually the aquatic world's record holders. (In case you're wondering where he finds barnacles in Alberta, he does his research at the Bamfield Marine Sciences Centre, located on a remote stretch of Vancouver Island's west coast.)

If you've ever been to the seaside, you've seen barnacles. They're the small shells, often less than a centimetre (a third of an inch) across, that encrust piers and coastal rocks the world over. They're crustaceans, which makes them related to shrimps and lobsters, but unlike their mobile cousins, they're glued in place. This creates a bit of a problem for the barnacle. While most anchored animals simply broadcast their sperm and eggs into the water, that's not an option for the barnacle. Thanks to their evolutionary history, barnacles fertilize their eggs internally. Which means they've had to come up with a way to get sperm from one barnacle at least as far as the next.

According to Mr. Neufeld, that means "evolving the world's longest penises." Yup, the longest in the world. At least, in relation to body size. This tiny creature's penis can be up to four times its body length when fully extended, which is a real advantage. After all, the longer the penis, the more mates it might be able to find in the area around where it's sitting. Or, as Mr. Neufeld puts it, "They take their penises and they extend them out into the water column, and they just simply sort of feel around until they find a potential mate. And if that mate is receptive, then they simply deposit sperm into the mantle cavity of that individual." And there are plenty of potential mates around for each barnacle. These sea creatures are hermaphrodites, so each one has a penis but will also produce eggs for another barnacle to fertilize.

You might be wondering what a barnacle penis looks like. First of all, when not in use, it's curled up in the shell, in what Mr. Neufeld describes as the barnacle's "armpit." They're ribbed structures, with hairs all the way along them, and another set of hairs at the tip, which Mr. Neufeld thinks they use to help them "grope around," looking for a potential mate.

If that's not weird enough, there's more to this story. Another feature of barnacles is that they have several legs that they can extend from under the top plates of their shell, legs used for feeding, not for moving. These legs are different lengths, depending on where in the ocean the barnacles are found. If the barnacle's home is in calm water, then the feeding legs will be long and feathery. But if it's on a stormy and wave-exposed shoreline, then the legs will be short and stout, robust enough not to get ripped off by the waves. Mr. Neufeld remembers the day when his lab decided to find out if this was also true of the barnacle's penis. "One day I walked into my graduate supervisor's office," he says, "and he simply said, 'I have two words for you. Barnacle penises.' And he gave me this look. To most people, this wouldn't really make a lot of sense, most likely. But knowing what we do about barnacle legs, I immediately grasped the implications of this: that if, in fact, they change the length of their legs, wouldn't that be cool if they could also change the size and shapes of their penises as well?"

To see if there were variations between penises from sheltered or exposed sites, Mr. Neufeld collected barnacles from various places, and then dissected them and measured the length of the penises. Being a good scientist, he also needed to know whether there was a relationship between the length of the flaccid penises and their "maximum extended length," as he puts it. "So we ended up taking off some penises, gluing them to a small capillary tube with crazy glue, and then artificially inflating them with a hypodermic syringe to see how far could they reach at their maximum extension."

Through these measurements, Mr. Nefeld and his colleagues were able to establish that the length of a barnacle's penis really is a function of where it's found. Those from the sheltered bays, where the water moves slowly, had the longest penises, while those from exposed rocks had a penis that was 20 per cent shorter and twice as muscular. "So," he says, "they're much more massive on the exposed coast, but much longer in the protected harbour."

This difference makes sense. If the barnacle is living on an exposed rock, it doesn't want its penis to get ripped off by the waves (sorry, guys, for that image), so it needs to be muscular and not too long. At the same time, if it's somewhere quiet, then having a longer penis will give it the advantage of farther reach, and more potential partners. It's a trade-off. The ones on the exposed coast are presumably growing the longest penis they can get away with, but the extra energy needed to make them more muscular will limit the length.

If the harbour were to become suddenly exposed, or the shore suddenly sheltered, the barnacles would be able to respond. In a final experiment, Mr. Neufeld transplanted barnacle larvae between the two different environments and found that the penis a barnacle grew reflected where it grew up, not where it was born. In other words, the length of the penis will adapt to the local conditions, and is not set genetically at birth. This is pretty astounding. Not many creatures can change how their body grows in as dramatic a fashion as the barnacle, with its legs and penis.

It certainly gives you something to think about the next time you scratch yourself on a barnacle when you're down by the sea.

PENIS PREFERENCE IN MOSQUITOFISH

Does size matter? It's certainly a stereotype that men are often accused of bragging about, and showing off, their male endowment. Women tend to laugh at this male obsession with size. But according to Dr. Brian Langerhans, currently a biologist and evolutionary ecologist at the University of Oklahoma, there are some females who particularly prefer a male with large genitals that he'll dangle in front of them. Female mosquitofish (*Gambusia* sp.), for example, are especially impressed. But for male mosquitofish, there are both risks and rewards for being well endowed, and it's the balance between them that matters.

The size of a male's genitals, called gonopodia in the case of fish, is a matter of serious study. Of all the visible body parts of a fish, it's the gonopodia that are the most variable. Two fish that are closely related can have the same colour, size, and body shape, but their gonopodia can be of dramatically different sizes. Which leads to the obvious question: why? If the primary function of the male sex organs is to transfer sperm from the male to the female, then the only factor that should affect its size is how well it functions to move sperm. The mosquitofish leads researchers to think there may be more to the story.

The different species of mosquitofish all look very much the same. They're small fish, only about two and a half centimetres (one inch) long. They're grey, and fairly nondescript. They bear live young, which means the male has to fertilize females internally. The most remarkable feature is the males' gonopodia. Dr. Langerhans says, "They have to have male genitalia that are non-retractable and hang off the belly of the body. And they have to insert it into the female body to fertilize her eggs." He adds, "Basically, it dangles from the belly of the fish if you look at the fish from the side. And they can swing it around wildly, swing it during courtship as well as during copulation." Depending on the species, the size of the gonopodia can vary greatly. Some species have an organ that's 20 per cent of the body length, while in others it can be up to 70 per cent. Guys, imagine dragging that around in the water with you.

Actually, dragging it around in the water is key to why some species have huge gonopodia, and others small ones. Mosquitofish live in many different environments, some of them populated with lots of predators, others with only a few. If you're in water with lots of predators around, the last thing you want is drag – you'll end up as lunch. So, you need a smaller organ. And that's exactly what researchers found. The mosquitofish that live in areas with lots of predators typically have much smaller gonopodia than those that live in predator-free places.

But why have a big organ at all, if a smaller one works? That's where the other half of this research comes in. Dr. Langerhans set up lab experiments where female fish could watch videos. (At the time, he was a master's student at Washington University.) "So the female in a little staging aquarium watched two videos of different males," he says. "Each video simply showed a male doing his courtship behaviour, which essentially means just swimming around and lowering his gonopodium and erecting his fins. But I digitally enlarged the gonopodium in one of the males by 15 per cent."

When he analyzed his results, Dr. Langerhans found that the females spent the most time close to the male with the larger organ. This presents a challenge for the male. A big gonopodium is going to help him get the female. But if it's too big, he won't be able to escape as fast when he's attacked by a predator.

This leads to trade-offs. In all environments, the male is going to have the largest gonopodium possible, as long as he can still escape predators. The fewer the number of things that will eat the fish, the bigger the gonopodia. Eventually, there will be a balance in the population, all determined by the environment.

For scientists, this finding opens up a whole new area of research. Not many have looked at why some species are better endowed than others, at least not in terms of environmental factors. If someone did, maybe she could finally answer the age-old question of whether size really matters.

Amazon Mollies

Sometimes nature just breaks all the rules. Take the Amazon molly (*Poecilia formosa*), for example. This tiny fish from Mexico and the southern United States, which measures three to seven centimetres (one to three inches) long, reproduces clonally. That means the female produces eggs that have all the genetic material the baby fish needs. No males are required. This also means that all the offspring are female too, exact clones of their molly mother. The

Amazon molly is an all-female species. But, in a bizarre twist, they still need to mate with a male to stimulate reproduction and produce offspring, even though they don't use the genes from the sperm. Which creates a dilemma. With no male Amazon mollies to mate with, they have to search out their lovers among other species. And, according to Dr. Andrea Aspbury, a biologist at Texas State University at San Marcos, that's where things get really complicated.

The Amazon mollies need to find a male who's willing to mate with them. He will have to be from a closely related species in order to be willing to mate. Then they'll discard his sperm. Exactly what this does to stimulate reproduction isn't known. But what is clear is that it's absolutely necessary. How does something like this come about? Dr. Aspbury explains, "These fish are of a hybrid origin. They formed from a hybridization event between the Atlantic molly and the sailfin molly at least 100,000 years ago. Something went on, such that the hybrid species that was formed no longer has sexual reproduction. They have this asexual reproduction."

Hybrids that aren't able to reproduce sexually wouldn't be that big a shock. That's what a mule is, after all (the hybrid of a horse and a donkey). But coming up with asexual reproduction, instead, is a bit of a jump. What it has done, though, is provide these females with a ready source of potential males to get sperm from. These mollies will mate with males from either the Atlantic or sailfin species in order to get reproduction going.

So what's in it for the male? At first glance, there doesn't appear to be much. But, says Dr. Aspbury, "Males might actually be able to increase their attractiveness to the females of their own species by a process called 'mate choice copying.' The idea here is, perhaps a sailfin molly male could increase his attractiveness to his own females if he's seen mating with another female, even if that other female is not the correct species."

It seems that female fish prefer males they've seen with another female. Perhaps that makes him seem virile and sexy in their eyes. Luckily for the Amazon mollies, they look very similar to the sailfin and Atlantic molly females – similar enough that humans can't tell them apart. So the females that the males really want to mate with are fooled into thinking their boys are hunks of burning love.

Even so, while mating with a female that looks like your type but is really a different species might make you look more attractive, it is more work for the male. So Dr. Aspbury took a look at whether there were any differences in male behaviour or physiology when they were with an Amazon molly female and when they were with their own kind.

The first thing she noted was that male mollies do prefer to hang around females of their own species. Amazon mollies are not their first choice. This makes sense. We might not be able to tell the difference between the females, but apparently the males can, and they adjust their behaviour accordingly.

More curious is what was happening physiologically. Dr. Aspbury measured the amount of sperm males produced when they mated with an Amazon molly or with one of their own species. She found that males produced up to twice as much sperm with a female of their own species. Not only that, but, says Dr. Aspbury, "Those males, in the presence of the Amazon molly, ended up decreasing sperm production relative to their initial baseline amounts of sperm." In other words, they're able to control how much sperm they produce, depending on which species of molly they encounter.

Can this last, long term? Probably not. Clones take a long time to adapt to changes, and unless there's something that is benefiting the male mollies, it's hard to see how this could persist into the deep future. But, for the time being, the Amazon molly is doing fine. And for males everywhere, it shows that while their genes might not be needed, at least the males are still good for something!

DOLPHIN BLING

We've all seen how young humans, when seeking the attention of a potential date, and possibly in an attempt to intimidate rivals, will use various objects to enhance their attractiveness. It might be a fast car, gold jewellery, or skimpy clothing. In the rest of the animal kingdom, though, enhancement through objects is rarely seen.

Most species get by with bright feathers, or large antlers, or a flashy mane of hair. But not always. There are some species that use tools for attraction. If you've been reading this book from the beginning, you've already met the bowerbird, which decorates its pad with cool knick-knacks. The Amazon pink river dolphin (*Inia geoffrensis*) is another example. This unusual cetacean species is a tool user, too. At least that's what Dr. Tony Martin, a vertebrate ecologist from the British Antarctic Survey, thinks after observing them looking for love with objects in hand (well, beak) for the last fifteen years.

Using a tool isn't the only thing about the Amazon pink river dolphins that's somewhat out of the ordinary. Take their colour, at least the colour of the males. The males really are, as the name suggests, pink. How they get that way is intriguing. Dr. Martin says, "What we've found in the last few years is that the pinkness is actually de-pigmentation. What happens is that the males fight with each other a lot, and the more they fight, the more they scratch each other with their teeth. And every time it gets scratched, an animal, when it heals, has a little bit more pink in it. And eventually, the older guys are completely pink. So they're almost luminously pink when they get to a ripe old age."

As well as being pink, these dolphins don't really look that much like what you and I picture when we think of dolphins. Because they live in rivers rather than the ocean, they need to be able to swim through shallow, dark water, which means they need a much more flexible body than their marine relatives. They also have big paddle-like flippers that allow them to swim backwards if they need to, and long, flexible jaws that allow them to pick up fish one at a time. Finally, they're almost blind. All this because they've been living in the Amazon, isolated for millions of years.

Dr. Martin actually discovered their use of tools by mistake. When he first started studying them in the Amazon, he noticed that they would pick up objects and throw them around. This wasn't much of a surprise, since dolphins in aquariums around the world are known to play with objects if they have access to them. So the assumption was that this was just another example of dolphins at play. But then, on a trip a few years ago, what Dr. Martin saw made him change his mind. "I saw an animal pick up a stick and go off with it," he says. "And it was flaunting it around. I just noted it down." The third time he saw this behaviour, he started to pay attention. He realized that it was only the males that were picking up objects and flaunting them. This made him think there might be another function to this behaviour besides play.

The objects come in all kinds of shapes. Sometimes it's a rock or a lump of clay from the riverbed. More often, though, it's something that floats. A stick, a bunch of leaves, or a bunch of weeds all make the grade. Not that this is a gift for a female. Dr. Martin says, "They seem to use it as a form of display, to show, 'Look at me! Look how great I am!' And sometimes in these groups, when this behaviour occurs, there are several different males all trying to outdo the others." It's a little like everyone driving a fast car, or wearing a gold necklace, when out to get noticed on a Saturday night. And, just as it can among humans on a night out, it can turn aggressive among these dolphins at times, but only when they're showing off their ornaments.

Unfortunately, a lot about how these objects are used remains a mystery. The water of the Amazon is extremely murky, so it's only when they bring stuff to the surface that researchers can see what's going on. What is known is that this behaviour occurs only when mating is common. This suggests that it's a sexual display. At least, Dr Martin assumes it is. "I can't prove that it leads to mating," he says, "because, in fact, we've never seen copulation in thousands and thousands of hours of observing this species. But we assume that they do copulate because they do make babies." It's also clear that the biggest males are the ones who show off the most objects, and that they're the ones who mate the most, which points to this being a part of the mating ritual. It seems, just like us, the river dolphins like to show off.

There's a serious side to this story, though. In the last few years, the number of Amazon pink river dolphins has dropped dramatically, for a variety of reasons connected to human contact. Let's hope we can control the number lost before they go extinct, and allow them to keep showing off their bling.

Cichlid Dominance

Imagine waking up one morning to find that you're the supreme ruler. You've not had to mount a bloody coup, or even an expensive election campaign. The current ruler has simply gone, and you've taken his place. That's the scenario researchers found was being played out in the competitive world of a species of cichlid fish found in Lake Tanganyika in Africa called *Astatotilapia burtoni*. Rising to dominance in this cichlid society usually involves a battle or two, but Dr. Sabrina Burmeister, a neuroscientist at the University of North Carolina, wondered what would happen if she simply took away the dominant fish. What she discovered revealed some surprising things about brains and behaviour.

The *Astatotilapia burtoni* are found in pools along the shores of the lake. They're intensely social little animals, living together in communities dominated by one male. Identifying that male is very easy. He's the largest and the brightest of the fish in the group, either a bright blue or a yellow, with a red patch and a stripe across his eyes, almost like a mask. The rest of the males in the group will be a dull, silvery colour, much like the females. The dominant male can be rather mean, so the rest of the fish stay out of his way, with the subordinate males all waiting for their chance to become the leader of the group.

In the wild, when that chance comes along, the formerly subordinate male will fight his way to the top, and then certain physical changes occur. "The male who wins dominance," says Dr. Burmeister, "will undergo both this behavioural transformation, where he starts behaving differently and he's brightly coloured, but also his fertility changes and his testes grow in size and he starts producing many more sperm. And that enhances his fertility and enables him to capitalize on his territory."

Since this change in the wild is usually the result of a fight, it's not easy to isolate when these subordinate males realize they've won, and now they are the top fish. So Dr. Burmeister wanted to come up with a way in the lab that would allow her to spot exactly when the male realized he was going to be in charge, and then find out what had happened in his brain. It wasn't an easy experiment to work out. Dr. Burmeister thought that measuring what was going on in the brain would be the difficult part, but, she says, "Designing the behavioural experiment was actually much trickier because these animals are very difficult to fool. We finally stumbled on a way of doing this. The night before, I went into the animal rooms where the fish were housed, in the dark with night-vision goggles, and I removed the dominant male. The other animals could not see this occurring. Then when the lights came on, the subordinate male was suddenly in a new situation."

The morning changes were astounding. Within the first two minutes of the lights going on, some of the subordinate males had started to change physically. They started to colour up, their eye bands appeared, and after twenty minutes, they looked fully dominant. The same was true for their behaviour. Even the least reactive of the subordinates had begun to change within thirteen minutes. "It appears to me," says Dr. Burmeister, "as I was watching them, that the males undergo a categorical shift, where they seem to recognize this opportunity, and then they go for it all the way."

This was interesting, but the key issue for Dr. Burmeister and her colleagues was what happened inside the fish's brain to trigger this change. They were able to find one gene, called EGR1, that seemed to switch on in response to the dominant fish's disappearance. This gene is important because it triggers a range of other genes to start working, which in turn release hormones and cause the changes in the appearance of the new leader of the pack. For Dr. Burmeister, this is special. "We're normally used to thinking that genes regulate our behaviour," she says. "In this model system in this species, what we can see is how behaviour regulates gene expression. We're showing that the behaviour of the animal in this social environment is regulating gene expression in his brain, which can then change his physiology and his fertility."

The influence of behaviour on genes is something that very few people have examined. Dr. Burmeister thinks there's room to look at this in humans, as well as in fish. "If we can understand how social environment can regulate these hormones in the brain that control fertility in fish," she says, "that can tell us something about how these same kinds of things are regulating these hormones in humans."

Who said there was something fishy about genetics?

3

PARENTING SKILLS

RAISING THE CHICKS

Birds put a lot of time and energy into raising their young, but sometimes chick rearing can take some strange turns, as these stories show...

COWBIRDS' BAD INFLUENCE

You have to admit it – the cowbird (*Molothrus ater*) is certainly clever, if somewhat cruel, when it comes to child rearing. Rather than go through the hard work and sleepless nights involved in raising chicks, they just drop off their eggs in another bird's nest and fly away. The unfortunate receiving bird, usually a songbird, is then stuck raising the large, demanding, and noisy cowbird chick. What's more, the songbird's own chicks have to deal with this lout in the nest, competing with them for their parents' attention. Dr. Liana Zanette, a biologist at the University of Western Ontario, wondered how the young songbirds could fight back against the interloper. What she discovered was who is really paying the cost of the intruder in the nest.

The cowbird is what is known as a brood parasite because of its habit of laying eggs in other birds' nests. Why the songbirds would put up with this is something of a mystery. According to Dr. Zanette, one hypothesis is that cowbirds are relative newcomers to eastern North America. That means the songbirds aren't yet used to them and don't recognize the strange eggs in their nests.

Another hypothesis is one that Dr. Zanette calls "mafia behaviour." Cowbirds will hang out close to the nest where they've laid their eggs. Then, if the songbird recognizes the strange egg and kicks it out of the nest, or damages it, the cowbird swoops in. What happens next isn't pretty. The cowbird will typically destroy the other bird's eggs, and even the nest itself, forcing the songbird to start over, building a new nest and laying another clutch. This suggests that those birds that don't destroy cowbird eggs actually have the evolutionary advantage, since their offspring get to survive.

But no matter which hypothesis is correct, the outcome is the same. The cowbird chicks can rely on the songbird parents for food and protection while they grow up. Of course, they still need to persuade their foster parents to feed them, so the cowbird chicks are loud. Dr. Zanette says, "They give these really loud ear-piercing screams. And so when a parent comes to the nest with food, it notices this individual and it thinks, 'Oh, my gosh, this is one hungry baby. I better feed this baby.' And it does so at the expense of everybody else."

This is not a great scenario for the other chicks. Not only is there an extra mouth for their parents to feed, but that mouth is bigger and louder and so is going to get the most attention. But song sparrow chicks (*Melospiza melodia*) have come up with a way to fight back. When Dr. Zanette compared song sparrow chicks in nests with and free from parasitizing cowbird chicks, she found something odd going on. In the unparasitized nests, she says, "What we found is that the sparrow chicks are pretty laid back. They give soft vocalizations, low pitch." Put a cowbird chick in there, though, and things get really noisy. According to Dr. Zanette, "When there's a song sparrow in the nest with a cowbird, it really ramps up its decibel level. It ramps up the pitch of the sparrow's vocalizations to such an extent that it resembles the cowbird chick a lot more than it does song sparrows in unparasitized nests. So it's behaving more like a cowbird than like members of its own species."

Dr. Zanette has a name for this behaviour. She calls it a "sheep-in-wolf's-clothing strategy." And it works. Parent birds feed everybody equally. At the end of the breeding season, the songbird chicks are as healthy as they would be if they hadn't had to share the nest with a cowbird. Talk about having a bad influence in the nest! Dr. Zanette was astounded when she saw the behaviour for the first time. "To me," she says, "it's really amazing that they could alter their behaviour to such an extent, even though they're just little guys. But you know, when they're in competition for survival, I guess they'll just do anything."

There's even more to this story. The baby songbirds are a canny bunch. They don't waste their energy on loud singing when they don't need to. The cowbird chick in the nest spends a lot of time chirping away at the top of its lungs, and this sends the nervous parents out hunting frantically for food to feed these obviously hungry mouths. But the songbird chicks hang back, only chirping occasionally – until the parents come back to the nest laden with lunch. That's when they'll ramp up the volume and compete with the cowbirds for the food. So not only are they competing successfully for food, but they're making the infiltrator do all the hard work. Sounds like a fair trade-off for having to share.

There is one great potential loser in this game, though, and that's the parents. It takes a lot more food to satisfy a nest that's got a cowbird chick in there, and while the young seem to be doing all right, it exhausts the parents. That's going to mean no second nest in the season, and fewer chicks over their lifetime. Or, if they do lay a second clutch, there will be fewer eggs. Either way, the parents lose out.

In this case, the kids really do seem to be driving their parents to an early grave.

THERE GOES THE NEIGHBOURHOOD

Seabird colonies make for a spectacular sight. Tens of thousands of birds, all crammed on to a cliff face, flying off to capture fish, and returning to the hungry mouths of their chicks. These colonies are crowded and noisy, but apart from occasional threats from predators, they are safe and friendly places – a good neighbourhood to bring up your chicks in. That's certainly been the case for a colony of common guillemots (*Uria aalge*, called murres in North America) on the Isle of May, off the coast of Scotland. That is, until the summer of 2007. In the space of one summer, what had been a peaceful, gentle colony became something much more sinister. Kate Ashbrook, a Ph.D. candidate at the University of Leeds in England, was there to witness the change.

It's amazing that these colonies ever live in harmony at all, as the birds live in very dense communities. There can be as many as thirty breeding pairs in one square metre (yard) of space, which means their nests will touch one another. And this is all happening on almost sheer cliff faces. Wherever there's a ledge, there will be nests three or four deep. There's not even room to build a proper nest; it's basically just a matter of laying an egg and holding on. This keeps the number of young down; pairs usually produce just one offspring. With all this close contact, and limited real estate, there's a lot of tension. Ms. Ashbrook says, "Normally, the adults are quite aggressive to each other over their space, so there can be a bit of fighting and wing beating between the adults. But there's never really been seen much aggression towards the chicks themselves."

You can imagine how hard it is to study these birds. The cliffs make the island itself difficult to access, and when you do, you have to lie down at the top of a cliff and peer over to watch the guillemots. Ms. Ashbrook says, "You start off at the beginning of the season and you know you're going to die if you fall off." Not a job for the faint of heart. But it does mean you can keep a quiet eye on these birds.

What Ms. Ashbrook saw in the summer of 2007 really surprised her. "We noticed," she says, "that breeding adults were actually attacking all the neighbours' chicks." This was something new. But then again, so was the behaviour of the parents. In a normal year, one parent will stay with the chick while the other goes off hunting for food. But not in 2007. That year, both parents were leaving the chick alone, sometimes for quite long periods of time (Ms. Ashbrook saw one chick left for sixteen hours). These chicks were left standing on the parents' breeding territory, a space only about five centimetres (two inches) across, which is smaller than the chick itself. This left the baby birds vulnerable to the weather, to gulls, and to neighbouring breeding pairs.

This wasn't the worst of it. When these youngsters were left alone, they would seek shelter from the wind and rain, which meant heading off the nest site to look for a sympathetic adult. Or, if a bird landed nearby with a fish, the chick might head over to investigate. And that's when the trouble really got bad. "Once off its site," says Ms. Ashbrook, "the chick was at high risk of being attacked by another adult. If a chick was attacked, it would run away from this first attacking adult and there would be a series of attacks by other birds that ran across the ledge. And then, sometimes, the chick was either thrown off the ledge or it was pecked so much it couldn't get back up again."

Ms. Ashbrook described it as "horrendous." Not that the adults were chasing down the chicks – they attacked them only when the chicks crossed onto their territories. It looked as if the chicks were after protection and the adults were avoiding taking on the responsibility of another mouth to feed. While this behaviour does happen occasionally in these colonies, the level in 2007 was way above normal. To give you a sense, in a normal year, the number of chicks lost would be about one in ten, but in 2007 two thirds of the chicks died.

What was behind this change? Ms. Ashbrook believes it was a problem with food resources. Why the number of fish would be down isn't clear, but it seems that both parents were having to spend more time than usual hunting. That meant they were both off the nesting ground at the same time, leaving the youngsters unprotected. The adults were noticeably thinner than in a normal year, and a hungry bird that's having trouble finding food for itself and its offspring isn't going to take well to others' children coming around begging. The upshot: a lot more aggression than is seen in a richer year.

What happens from here is unclear. Without knowing why there were problems with food collection, it's difficult to say whether this was an anomaly or the start of a growing trend. If it was just an abnormal year, the colony will probably recuperate, since these birds are long-lived and will continue to breed and rebuild the year's losses. If, on the other hand, the problem is collapsing fish stocks, then we may have seen the beginning of the end for the guillemots of the Isle of May.

Birds that Time-share

Here's a bargain for you: a beautiful time-share on a gorgeous tropical island. You just have to be willing to share it with a bunch of birds. That's not the typical pitch you get for a time-share, and you're probably thinking that sharing with birds wouldn't be so great. But these birds are more than willing to share. They do so every year. Dr. Vicki Friesen, a biologist from Queen's University in Kingston, Ontario, has been studying the band-rumped storm petrel (*Oceanodroma castro*) and has found that not only do they have this weird behaviour of co-habitation, but their lifestyle may prove a 150-year-old hypothesis originally put forward by Charles Darwin.

Most people have probably never seen a band-rumped storm petrel. They're small birds, weighing in at about forty-five grams (one and a half ounces), roughly the size of a songbird. They're a dark grey colour, with a white band at the top of the tail (hence their name), and they have a habit of pattering their feet on the water when they feed. This last characteristic helps explain the rest of their name. Petrels are named after St. Peter, who is said to have walked on water, which is what this pattering action looks like. And for most of the year, these birds spend their life out at sea.

Once a year, though, they come in to islands to breed. They don't stay too long – land is dangerous for seabirds, since there are plenty of predators around. So far, nothing very unusual. But there is one group that shows the odd co-habiting behaviour mentioned. Dr. Friesen has studied these birds on a number of islands around the world in tropical and subtropical areas. "In five archipelagos," says Dr. Friesen, "there's one group of storm petrels that will come in and nest in burrows. And when they're done laying their eggs and raising their chicks, and they go off to sea, then another set of storm petrels comes in, and they lay their eggs and raise their chicks often in the very same burrows. And then they go off to sea, and the first set comes back again and repeats the cycle." It's the perfect time-share. Neither group bothers the other, and both have a place to stay exactly when they need it.

That's not the only interesting thing about this storm petrel. The other interesting aspect has to do with speciation, the process by which separate species form. From a biologist's point of view, speciation has happened when two related animals can no longer interbreed. Or, if they do interbreed, the offspring are infertile (for example, a horse and a donkey breeding to produce a mule). But how does one species split and become two? There are a few different hypotheses around; the most common describes a process called allopatric speciation. Here's a quick etymology lesson: "allo" means "different" and "patric" means "homeland." So, in allopatric speciation, a species gets split by geography or the environment. The two groups end up with different homelands. Maybe one group is stuck on one side of a mountain range, while the rest is on the other side. The two groups no longer intermingle, and over time, they evolve in different ways, so that eventually they can't breed together, or they don't recognize one another, and two species are born.

Darwin suggested another model: sympatric speciation. "Sym," in this case, means "the same." He thought that it was possible for the two species to have evolved in the same place, meaning there's no physical barrier to reproduction like the mountain range, but some other reason keeps the two populations apart. What would be the barrier? Darwin didn't say, and after his death, when genetics came to the forefront of biology, sympatric speciation was largely dropped as a possibility. After all, once you understand genetics, sympatric speciation doesn't seem likely. That's because if two groups of the same species share a physical location, any genetic exchange (a polite term for mating) will prevent speciation from getting started. No one could come up with a barrier strong enough to allow sympatric speciation to happen.

But these petrels have given us an example of how this happens. In their case, the barrier is time. Two populations that breed at different times of year, and don't spend any time on the island together, start to move apart genetically. That's what Dr. Friesen found was happening. The process of speciation is not yet complete in these petrels, but the two populations are indeed genetically distinct, and depending on which group you look at, they're more or less similar to the birds that share their nests. In the Galapagos Islands, the different breeding groups still have a similar genetic makeup, but in the Azores, the populations, even though they look similar, have very different sets of genes, different enough that they should be considered separate species.

It's a great example of evolution in action, and justification of Darwin's ideas. What's left is to figure out why it is happening. Dr. Friesen does have some ideas. "There are a couple of possibilities," she says. "Probably the simplest and the most popular one is that there was competition for either food or nest burrows on the island. So there probably wasn't a sudden switch of a bunch of birds from one season to the other, but, over time, young birds may have found that the feeding was better in a different season. And so they switched their reproduction to be in a different season from the rest of the population."

Now there's a response to a parent who's crowding you out. Go off and develop your own species.

Two-Mom Albatross Families

When the book *Heather Has Two Mommies* was first published in the late 1980s, it caused quite an uproar. It was the first children's book that addressed the subject of same-sex parenting. Much of the outcry against the book came from people who thought it was unnatural for a family to have two mothers. Apparently they didn't know that nature is just fine with the concept. Lindsay Young, a doctoral student in zoology at the University of Hawaii, has been studying the nesting behaviour of the Laysan albatross (*Phoebastria immutabilis*), and has discovered that having two mothers is the norm for these giant seabirds.

The Laysan albatross ranges all across the north Pacific; you can sometimes find them flying off the west coast of North America. They nest in the Hawaiian islands but come up the coast as far as Alaska to hunt for food. That's quite the commute to find lunch. In appearance, they aren't particularly remarkable. Ms. Young says they are "kind of like oversized seagulls, in a way." Then she adds, "I hate to make that comparison because they're much more graceful than those guys. They have black on the upper parts of their wings, and then the rest of their body is white. Similar to a gull. And they have a large, long, yellowish bill." This description applies equally to males and females. Unlike many bird species, both sexes look identical, and the only way to tell them apart is by examining their genes.

When Ms. Young first started looking at the Laysan albatrosses while they were breeding, she had no idea that there were families with two mothers. Her work was about what you'd expect from a study of breeding birds. First, she put bands around the different individuals' legs, so that she could tell them apart. Then she watched for who settled down with whom, whether or not their egg hatched, and then whether the chick survived long enough to make it out of the nest, and so on. But there was one big surprise. Some nests had two eggs in them.

This didn't make sense to Ms. Young. Raising an albatross chick is difficult. The adults have to travel far and wide just to bring back enough food to feed the youngster. Feeding two is too much for a pair of birds, so there's never a successful nest that fledges two young. Invariably, what always happens is that shortly after the two eggs are laid, one of them will fall out of the nest, and the birds won't continue to incubate it. That way, the surviving egg gets all the attention, and the one chick often does well.

When the research study into the Laysan albatross started, about 15 per cent of the nests contained two eggs. But after a few years, the researchers were finding that 25 per cent of the nests had two eggs. This prompted Ms. Young to look at the genes of all the adults, using small blood samples, to figure out what was going on.

To her surprise, she found that about 30 per cent of the two-egg nests were being attended to by two female albatrosses. "The first thing that went through my mind," says Ms. Young, "was, 'Wow, I really screwed this up.' And so I went back and actually did it four different times to make sure that this wasn't a mistake."

It wasn't a mistake; a third of the pairs really were made up of two females. Which, on reflection, made perfect sense. There's a severe shortage of male birds in this colony, with three females for every two males. And if a female can't find a mate, then she can't raise her egg alone. It makes sense for her to partner up with another female, and at least have a shot at raising a chick.

Which raises the question of who the fathers are. And the answer is that male albatrosses cheat on their partners. While all albatrosses form long-term partnerships, some of the males will sneak off and mate with other females, allowing them to produce viable eggs.

The pairings between females seem quite stable. By looking back over records of the colony, Ms. Young was able to establish that at least one pair of females had been together for nineteen years. Again, there's a logic to this. In a given year, only one egg makes it. So, if you're paired with another female who has also laid an egg, you can't guarantee that your egg is going to be the one to survive. When Ms. Young started to look carefully at this, it became apparent that there was no rhyme or reason as to which of the two eggs survived. So, just based on statistics, the longer a bird stays with her partner, the greater the number of her eggs that will survive. The energy involved in forming a new partnership, plus the risk of loss to her own egg, mean that a female albatross is better off staying in the relationship she already has.

Among these birds, jealousy isn't an issue when a female is incubating another bird's egg. Albatross aren't the smartest of birds out there. "Albatross are not good at telling their own eggs apart," says Ms. Young. "They've been known to incubate volleyballs, beer bottles, other birds' eggs." So it's better that they're at least incubating an egg, rather than something else.

One more thing for the morally outraged to consider. It looks as though this behaviour would never have arisen if it hadn't been for humans. The lack of males in this colony is the result of human colonization. When humans settled in Hawaii, they drove a lot of the seabird colonies away. It's only in the last few years that the birds have started to return, and, as luck would have it, females have led the way.

So the situation may change in the future, but for now, two moms is better than one, at least for the Laysan albatross.

Bringing up a Brood

Beyond the birds, there are plenty of other caring parents out there in the animal kingdom. Even insects will sometimes look after their young. But each creature faces unique challenges when it comes to being a parent.

Self-Roasting Rodents

You would think that the last thing a tiny squirrel living in the Yukon would have to worry about is the heat, but that's exactly what the problem is for the Kluane red squirrel (*Tamiasciurus hudsonicus*). Even when the temperature is below freezing, these little rodents have to work to avoid overheating. This is especially true for mothers raising young pups. The squirrels provide their own central heating. But, as Dr. Murray Humphries, a wildlife biologist at McGill University, has discovered, their thermostats are often stuck on high, which turns out to be too hot.

These squirrels need a very high metabolism. Particularly when they're lactating, the females need to get enough energy to not only keep themselves going but to nurture their offspring as well. That's true of all mammals, but these squirrels can have as many as four pups in a litter, and their combined weight can match their mother's. She really is eating for two. She's munching all day, then dashing back to the nest to feed her pups. But she's burning this energy off, as well. According to Dr. Humphries, in comparison to a human, "[Metabolically], red squirrels are running about the same as, say, Lance Armstrong would be during Le Tour de France." Here's the problem, though: if your metabolism is running like this, you're also going to be generating a lot of body heat. Ask any athlete and she'll confirm this. Sometimes this can interfere with your life.

That's the case with these red squirrels. If the females get too hot when they are producing milk, their milk production drops noticeably. In the lab, this has been tested by shaving females. The shaved ones will produce more milk than hair-covered squirrels. (As an aside, the same has been shown to be true of laboratory mice, and is probably true for most hair-covered mammals. The issue for red squirrels is that because they live in such a cold environment, they have a tougher battle keeping a hot/cold balance than species from a more temperate climate.) In nature, however, shaving is not an option for squirrels. So, Dr. Humphries says, "You'd expect them to not pick the most insulated nest out in the wild, but rather to pick ones that are kind of awfully insulated. Warm enough to keep their young warm, but not so warm that they would be limited by the heat they can dissipate." Call it the Goldilocks approach, finding a nest that is just right.

Except, what's perfect changes as the season wears on. At the beginning of spring, the pups are tiny and naked, and the outside temperature is quite low. So the mothers need to be able to keep everyone toasty warm. Then, later, the pups have hair, are larger, and the air outside is much warmer. Everyone is generating lots of body heat, and it's then that it's very easy for the mothers to overheat.

The squirrels do have a solution: moving around. They don't just set up one nest, they'll have several every season. Dr. Humphries says, "They sometimes nest in cavities; sometimes they nest in grass nests out on a tree branch; sometimes they nest on witches' brooms within the spruce trees." (In case you think he's pulling your leg, we should explain that witches' brooms are tree branches that have been infected by a rust fungus.) During his studies, Dr. Humphries and his students would measure the insulating characteristics of different nests occupied by one female. They would chase the female out of one nest (humanely, of course), and then put a bottle of warm water in the nest and note how slowly it cooled down compared to one left outside.

What they found was that, first of all, it didn't matter much where the nests were located; at the beginning of the season, they all had about the same ability to insulate. But, as the season went on, the nests females were using were less insulated than the ones they'd used earlier in the year. Not only that, but the amount of insulation changed depending on the size of the litter. The more offspring and therefore the more body heat around, the less insulated the nest would be.

There are two ways this could be happening. It may be that the females are choosing different nests in the later spring. Or they may be throwing out the insulation in already existing nests, to cool them down. Which of these is going on isn't clear yet. Unlike humans, these creatures actually want drafty homes!

Females have even been seen quite late in spring with babies in their mouths, wandering from nest to nest. This seems pretty inefficient. But constantly moving may help them deal with parasites, too, since fleas are a real nuisance for these animals, and if they keep moving the babies, then the overall parasite load will be lower. Still, it takes a lot of energy to keep carrying your kids from place to place. Dr. Humphries thinks this may be an evolutionary leftover. Being warm-blooded provides a real benefit to the offspring, giving them a great head start in life. Birds do this by sitting on their eggs. Mammals do this partially by developing their offspring in a womb. It naturally follows that they want to keep them warm while they're growing. But, says Dr. Humphries, "There's a kind of double-edged sword there. And it's quite easy to switch over to a point where now you've actually got too much heat in the system and you need to find ways to dissipate it." Which is what the squirrel does by switching nests.

So, if you're ever out showing a home to a squirrel, remember, drafts are in – if it's late spring.

FATHER'S NOSE BEST

For most baby primates, there's nothing quite as soothing as a cuddle from your mom. Well, if you're a human baby who's being cared for by your stay-at-home dad, a cuddle from him also does the trick. Just as it does for baby common marmosets (*Callithrix jacchus*). Male marmosets are well beyond all that chest-thumping, macho behaviour that's so popular among other primates. They're one of the few monkey species that shares parenting responsibilities, with the male often taking on the bulk of the care-giving. Dr. Toni Ziegler, a primatologist with the National Primate Research Center in Madison, Wisconsin, has been looking at marmoset fathers to figure out why they have such a tender touch. And she's discovered that when it comes to being a devoted dad, a marmoset father's nose knows best.

Common marmosets are found mainly in the forests of Brazil. They're about the size of a squirrel, weighing in at about 400 grams (14 ounces). They're also pretty easy to recognize. They're mostly grey, with a long tail, and they sport big white tufts of hair on their ears. In fact, some people call them cotton-eared marmosets because of this distinctive marking. However, it's not what they look like that's got Dr. Ziegler's attention, it's the fact that they live in family groups.

Having both parents raise their young is relatively rare among primates. We do it, and so do some species of gibbons. But among other monkey and ape species, the task of looking after junior usually falls to the mother, whether the father is around or not. A marmoset dad, though, will happily carry the kids around on his back. He also stimulates the infants to urinate and defecate, a skill they don't develop for themselves until after the first eight weeks of life. And while he can't nurse his offspring, as they get older, he does help them out. It's the father who teaches the young where to find food, and he will share his food with them until they can gather enough of their own. So his involvement is very hands-on.

But, as any parent will tell you, it takes a lot of patience to raise children. And, biologically, this caring role goes against the norm for males in the wild, where aggression towards rivals, and sometimes towards the female, is key to finding and keeping a mate. So Dr. Ziegler wanted to know how the fathers, who had been busy defending their mate choices before the young were born, were able to dial their aggressive instinct down once babies were on the scene.

Dr. Ziegler wondered if it might have something to do with the way the offspring smelled. To test this, she took several fathers out of sight of their offspring. Don't worry, not permanently, just for about ten minutes, in order to clear their nostrils of any scents from their family. One at a time, she let each male smell a piece of wood laced with the scent of his child. Then she waited another ten minutes before taking a blood sample. Her idea was to see whether there were any changes to his blood chemistry after smelling his infant, compared to when she performed the same experiment with a blank piece of wood.

The results were curious. "We found," she says, "that when we looked at their testosterone levels compared to the control, they had a significant decrease in testosterone after smelling their infant's scent. We also tested males that were not fathers, and they showed no difference in testosterone levels between their control and an infant scent."

This lowering testosterone may help explain the patience of the parenting male. It's well established that testosterone is involved in aggressive behaviour. For instance, we see rises in testosterone levels when a male is involved in fighting or mating. Conversely, the lower the testosterone level, the less the aggression, making a male more tolerant of the demands of a youngster.

What's especially interesting about what Dr. Ziegler discovered is how much the environment influences hormone levels. Just a few moments of exposure to the smell of an infant, and a male's brain is sending the message to produce less testosterone. And it's the smell alone that has this effect. Remember, the males couldn't see their children during the experiment. While it's clear that environment has an effect on testosterone in general, it's a surprise to see the hormone system reacting so quickly and fluidly to change.

Dr. Ziegler thinks that marmosets might not be alone in responding to the smell of an infant. "One of the things that is known about humans," she says, "is that fathers actually have lower testosterone when they have young children. It's never been looked at as a cause-and-effect, but generally they're shown to have a lower level of testosterone. So it could be the same system that's occurring."

After all, monkeys aren't the only primates where father's nose best.

Pregnancy and Monkey Dads

We all know that expectant dads often do a lot at home in anticipation of a new baby arriving. Perhaps they build a new cradle, or paint the nursery, or make late-night runs to the corner store to satisfy the cravings of their mate. But some primate males take their preparations for fatherhood one step further. In her study of common marmosets (*Callithrix jacchus*) and cotton-top tamarins (*Saguinus oedipus*), Dr. Toni Ziegler, the primatologist from the University of Wisconsin-Madison whom we met in the previous story, has found that the males pack on weight when their partners are pregnant.

This all started as a simple observation. Dr. Ziegler and her colleagues have been studying monkeys for a long time and had noticed that, as she says, "along the breast area, they looked like they had maybe some fat pads or something." This gave the team the idea of weighing the fathers to see how much they gained. "We weighed the fathers monthly during the entire five months of pregnancy in the common marmoset, and six months of pregnancy in the tamarind, and the females were weighed monthly as well."

Their eyes hadn't been deceiving them. The fathers really had gained weight. Even before the females had started to put on the ounces, which happens in both species during the last couple of months of pregnancy, the males had started to plump up. Not insignificantly, either – the males were gaining anywhere from 2 to 15 per cent of their body mass. Luckily for the body conscious among them, these monkeys are quite hairy, so it's not immediately obvious that they're heftier. But there's more of these big boys to hang on to when their mates give birth.

And hang on is what the babies do. After the infants are born – and both species typically produce twins – it's mostly the father who carries them around until they can walk on their own. Considering that he's generally toting two babies at a time through the trees (the babies together can weigh up to 20 per cent of his body mass), he's going to need a lot of energy. Which is what

Dr. Ziegler thinks is behind this weight gain. While the infants are small, using the extra fat helps him sustain his energy levels enough to keep them safe.

The story behind what's triggering this weight gain in the males is complicated. Dr. Ziegler has tested the hormonal changes in the males and found that they start producing lots of prolactin. This hormone stimulates milk production in females, but is also linked to weight gain in both sexes. What prompts this increase in prolactin in the males is less clear. What's known is that it happens just after another hormone called cortisol spikes in the females. This, in turn, is stimulated by the growth of the fetus. How this rise in cortisol levels in the expectant mother affects the prolactin levels in the father isn't yet known.

There are a couple of possibilities. It could be that the father is picking up the hormone in something he's eating. Another option is that he's taking it in through his nose. Monkeys have a highly developed organ in the nose for picking up pheromones. If the cortisol levels in the female are causing her to release some pheromone, then her mate is going to smell it. Up go the prolactin levels, and on goes the weight.

There's a second possible benefit, beyond weight gain, to these hormone changes in the male. Many hormones change behaviour as well as physiology, and, according to Dr. Ziegler, "some of them actually cause responsiveness to infants. So I think that the enhancing of these hormones is actually preparing the father for this investment in his offspring."

If this looks like a sympathetic pregnancy to you, well, it does to Dr. Ziegler as well. Which brings up an interesting human parallel. There's a condition among men called couvade syndrome, in which men exhibit some of the same symptoms of pregnancy as their spouses. While no systematic research has yet been done on the subject, this work hints that there may be something to it.

Either that, or human males can just join their mates in scoffing down the ice cream late at night.

STINKING BABY BUGS

"I'm hungry!" Every parent has heard the whiny appeal of a child who wants to eat, *right now*. And any parent can tell you that begging is a pretty effective tactic. Just try to ignore a child's cries for any length of time. But what if you're an infant who can't call out? How do you let your mom know that you're hungry if you have no vocal cords? That's the problem faced by baby burrower bugs (*Sehirus cinctus*), and so they turn to another of their mother's senses. Dr. Butch Brodie, a biologist now at the University of Virginia, has worked out which of the senses that is, and how the baby bugs communicate with their insect parents.

The burrower bug is no poster child of the insect world. You've probably never seen one, and even if you have, you probably didn't pay it much attention. The adults are small, shiny, and black, about the size of a pencil eraser. The offspring are about the size of a pinhead, and are more interesting to look at because they're bright red. But beyond that, there's nothing particularly remarkable about their looks. They are quite common, though, in southern Canada and throughout the United States. Dr. Brodie finds them in the early spring, before ploughing, in corn fields, where they live under leaves and the litter on the ground. They wander around a lot, looking for mint seeds, which they then eat.

Dr. Brodie finds them interesting because they're insects that provide care to their offspring. The mothers produce between 100 and 150 eggs at a time, and then feed them while the larvae develop. It's a model system for the study of parental care. And what researchers are looking for when they study systems of parental care is how the parents' conflicting needs are juggled. The parents wants their offspring to survive, but at the same time want to minimize the energy expended raising them. As Dr. Brodie says, "Invariably, it's the case that kids want more resources than parents are willing to give, and so in all sorts of animals, you've got interactions that result in kids manipulating the care level that parents provide."

That manipulation, in many species, involves sound. This would be the begging cry mentioned earlier. But in earlier experiments Dr. Brodie had been able to show that sound made no difference to these insect mothers. This time, he decided to look at smell. After all, most insects communicate information through smell.

Testing the idea that the young might be chemically communicating with their mothers required some invention. Members of Dr. Brodie's lab created a device he named the "smell-o-tron." "It was an elaborate set of glass tubes," he says, "where we could put babies into the glass tubes, blow air over them, and then expose other mothers to the odours of kids that had different manipulative condition levels." (He means whether or not they had been fed recently, so the mothers could tell how hungry they were.)

Success! When Dr. Brodie placed hungry, underfed infant bugs in the tubes and blew air over them, their moms responded by bringing food for them, as much as three times the amount they brought when they detected the scent of well-fed larvae. He seems to have found evidence for chemical signalling between parent and child.

The chemical makeup of that signal is still something of a mystery, although Dr. Brodie can identify several specific components. For instance, there's alpha-pinene and camphene in there, but when Dr. Brodie tried to trick mothers into feeding their offspring by exposing them to just these two smells, nothing happened. What he thinks is important to the mothers is the ratio of these different chemicals, which changes depending on the hunger level of the little ones.

If you're thinking of taking a sniff of these bugs, be warned that they're hard to detect this way. They are very small, remember, but if you do manage to get close enough, you'll detect a hint of pine.

There's a great deal left to learn about these bugs. How the babies change the levels of their smell, for instance. Or why the mothers respond so strongly. One thing is clear, though: theirs is a very unusual system. This is the only chemical-based system for communicating hunger that's been discovered. Until now, research has focused on the whiny verbal versions. But if more insects are studied, we might find it's more common than we think. If that's the case, then sound might play a role in fooling the parents of pests, and preventing their offspring from growing.

But for now, among the burrower bugs, it's the smelly babies that get the most food.

CANNIBAL TADPOLES

When you have kids and you're looking for a new home, one consideration is whether there are other children already in the neighbourhood. If there are, there will be lots of playmates for the little ones, parents with watchful eyes, and probably good schools. Things are a little different, though, for the Peruvian poison frog (*Dendrobates variabilis*). When these frogs are looking for a home for their tadpoles, they look for places where other eggs have been laid, but they're not hoping these eggs will grow up to be playmates for their young. No, they're hoping these eggs will grow up to be food. Dr. Kyle Summers, a biologist and evolutionary ecologist at East Carolina University in Greenville,

North Carolina, was the discoverer of this tadpole cannibalism.

These poison frogs are found in the Cainarachi Valley of Peru. They're quite small, about two centimetres (three quarters of an inch) long, and brightly coloured. They also have poison in their skin to protect themselves from predators. The bright colours are supposed to warn other animals off. In this species, the body is a metallic green-gold with large black spots, and the legs are green with smaller black spots. This little creature isn't trying to be inconspicuous.

These frogs lay their eggs around small pools of water, which collect in the centre of bromeliads, the showy, flowering plants of the rain forest. They don't lay their eggs right in the pool of water (which are just a few millilitres or ounces); rather, they stick them to the plant just above the pool, so that when the tadpoles hatch they drop down into the water. This keeps the eggs moist, but stops them from drowning.

After the eggs hatch, the male frog sits on the clutch. After the tadpoles hatch and drop into the water, they try to wiggle up on his back. The father will then take them, one by one, and as many as he can, to another bromeliad pool. That becomes the tadpoles' new home, and, as mentioned, the father will usually try to find a pool where there are eggs.

All this is done for the sake of food. It's been known for a while that larger tadpoles will eat smaller ones – even, in certain cases, tadpole eggs. The reason lies in the immediate environment. These pools of water are small and don't contain many nutrients. The only real source of food are the tadpoles themselves. When a tadpole hatches and slides into the water, it's at risk of being eaten by any larger tadpole already in the pool. Even if Dad comes along to take it away to another pool, where there are no larger tadpoles, there's a chance the baby may slide back in. And he can only take so many away before the babies start to get eaten.

Was this all just an accident, though – a coincidence that the males were moving their offspring to pools with a ready source of nutrition about to hatch? That's what Dr. Summers wanted to investigate.

He started out by setting up fake pools. These were just plastic cups with leaves wrapped around them, but they were good enough to fool the frogs. The mothers came along and laid eggs, and then Dr. Summers's team sprang into action. They took the cups and set them up in pairs: one with eggs, one without. Then they waited to see what happened when other frogs came along. Sure enough, the frogs would pick the pool with eggs. They were actively choosing to put their tadpoles where there would soon be food available.

"Talk about the school bully," says Dr. Summers. "This is the school cannibal. Imagine it from the perspective of the larvae in the clutch. It's been sitting, minding its own business, and all of a sudden it ends up with this giant tadpole below it, just waiting for it to slip up."

From a species point of view, this doesn't seem to make a lot of sense. After all, one tadpole might survive to become an adult, but only at the cost of many others that may be distantly related to it. It's probably some kind of trade-off. Growing up is hard in this environment, with not enough food to go around. So, as long as some make it through to become frogs, the rest of the clutch is expendable. Obviously this strategy works, as the species has been around for millennia. And it's better for the tadpoles than being moved to a big pool. Those are going to be full of animals and insects that would love to have a tadpole for dinner. In the end, it's better for the survival of the species for the tadpoles to be eaten by one of their own than to lose everyone to a predator.

4

LUNCH, ANYONE?

Finding Food

Gathering a meal is something every creature has to do. But sometimes, there are creative and innovative approaches that make going down to the corner store seem very pedestrian...

Owl Dung Nests

Seems you can barely turn on your TV these days without coming across another home renovation or decorating show. Whether it's *Extravagant Makeovers* or *Designs on a Budget*, every possible adornment has been trotted out – except perhaps the one that would appeal most to burrowing owls (*Athene cunicularia*). These birds of the field are themselves avid avian decorators, but it's unlikely you'll find one of their favourite ornaments on *Trading Places* or in the pages of *Better Homes and Gardens*. That's because burrowing owls like to line their nests with animal dung.

Dr. Doug Levey, an evolutionary ecologist from the University of Florida, thinks there's more than just decoration on the minds of the birds. He thinks they're playing with poop for a very practical reason.

Dr. Levey didn't initially know what that reason was, and it was only by accident that he even noticed the dung in the dens. "I teach an avian biology course at the university here," he says, "and we were out with the students one weekend to look at these owls. When we walked up to their burrows, the students noticed that there was dung all over the place, so they asked the obvious question: 'Why is that dung there?' And, heck, I didn't know."

Once he started looking into it, he found that dung is very common in owl burrows. The burrows themselves are about three metres (three yards) long, and run up to a metre (yard) under the surface. The owls bring back all kinds of dung, primarily horse and cow, but closer to urban areas he's also found dog and cat dung in the burrows. Basically, they'll take any dung they can find.

You'd think this would make the nests somewhat smelly, and Dr. Levey wondered if this might be the point of their using the dung. Burrowing owls are at risk from a number of predators, and maybe if their nests smelled of another animal, that would disguise them. But that turned out not to be the case. Dr. Levey set up artificial burrows, with quail eggs at the bottom, and half covered them in dung. When predators were given the chance to hunt the eggs in the smelly and clean nests, he says, "What we found was, there was no difference at all."

But there was a clue to the real purpose of their using the dung in that original observation with the students. Where the students had found feces, they'd also found owl pellets that were full of parts of dung beetles. Owls regurgitate pellets made up of the hard parts of their prey that they can't digest. "And so," says Dr. Levey, "we thought that perhaps this dung might be serving as bait to attract dung beetles. So we did an experiment where we cleaned off the aprons of all the owls' burrows in two colonies, and half of the owls we gave cow dung. The other half didn't get any. Then we left them alone for four days. At the end of that, we came back and collected all the pellets and prey parts from around their burrows. And what we found was dramatic. When owls had dung in front of their burrows, they ate ten times as many dung beetles and six times as many dung beetle species as owls that did not have dung in front of their burrows."

The owls are baiting their nests. They're getting their lunch to come to them, rather than wasting energy on hunting. It fits with what we know about owl behaviour. Unlike most owl species, burrowing owls are awake during the day, and they're often seen standing by the entrance to their nests. Dr. Levey's work shows that they're not so much protecting their burrows as waiting for dung beetles to come in – and then pouncing.

Before you start thinking that this is one really smart animal, there's something you should know. Using dung as bait probably started as an accident. Burrowing owls bring all kinds of things back to their burrows, particularly in the spring and early summer, including bits of plastic and aluminum foil. "Our favourite was two-dimensional, mummified toads that we think they peeled up off the roads nearby," says Dr. Levey. "And it wasn't just one pair of owls that would do this; we found it several times." Why they do this is a mystery, but it could be evidence that the bringing of dung was an accident that led to an evolutionary advantage, and a behaviour that spread through the population over time.

Even if it started by accident, it is an example of tool use in the animal world. Other birds do this too. Herons use bread for bait, for example. And it may serve other purposes. Could it be that a particularly smelly den is attractive to potential mates? That's a question yet to be answered.

And if dung is used to attract beetles, what will you get if you lay out a squashed frog?

SNAILS AND THEIR SLIME

When the mighty hunter sets out in search of prey, he needs certain key skills: stealth, perseverance, and, in the case of the rosy wolf snail (*Euglandina rosea*), a keen appreciation of slime. These greedy gastropods are tenacious trackers, chasing down other snails and slugs for lunch. Not that every snail they meet is an enemy. Sometimes it's another wolf snail that's on the trail ahead, and then it's all about dating, not dining. How do they tell whether they're on the trail of prey or play? That's the question Dr. Melissa Harrington, a neurobiologist from Delaware State University, set out to answer. She thinks it's all in the slime.

Native to the southern United States, but invasive in other tropical regions, the rosy wolf snail is quite large, at about ten centimetres (four inches) in shell length. It gets its name partly from its colour, which is a rosy brown, and the wolf part because it lives on a diet of other snails. It's sometimes called a cannibal snail, which Dr. Harrington says is a misnomer – though they are carnivorous, wolf snails never eat their own species. The most distinctive feature of these snails is a pair of what are called lip tentacles. These aren't the eye stalks we normally picture on snails. Instead, imagine one of those huge moustaches from the silent movies, the kind that curl up at the ends, ready for the villain to twirl. Except these lip tentacles aren't made of hair, they're the same flesh as the rest of the snail's lip. And the tentacles constantly move around, tasting the ground in front of the snail, in much the same way that a snake's tongue tastes the air.

What the wolf snail is looking for is another snail's slime trail. Yes, that messy trail that we see glistening on stone around our prize roses is what the wolf snail will follow. When it comes across a trail, the wolf snail will change direction to follow the slime. At this point, it is important that the snail goes the right way. After all, the snail that laid the trail will have gone in one direction, and if the wolf snail goes the opposite way, then lunch – or a potential mate – has got away.

But how does it know which way to go? The short answer is that the wolf snail doesn't necessarily get it right when it's hunting for food. In experiments with trails in her lab, Dr. Harrington found that the wolf snail went the right way, towards its prey, only slightly more often than chance would allow. But the same is not true if they're after a mate. When a wolf snail comes across the trail of another wolf snail, it will follow the potential suitor every single time.

The snails must be telling their left from their right so that they can orient in the same direction as the snail they're following. The problem is that Dr. Harrington has looked very carefully at the slime trails and can't find any clues. She says, "On a microscopic level, you can't even tell which is the long direction of the trail and which is the side-to-side direction. You get totally disoriented."

Her best guess has to do with the anatomy of snails. "Because of their shell," she says, "the right and left sides of snails are not the same. So snails, to use a highly technical term here, actually poop out of their neck on the right side and only the right side. So, given that they have this asymmetrical body plan, we thought it's possible that the trail might have a right and left side to it because the right side of a snail is different from its left." This isn't the snail's only asymmetry. They also have all their reproductive organs on one side of their body. To put not too fine a point on it, if they're pooping out of the right side of their body or if their genitals emit any scent, then there may be more of certain chemicals on one side of the trail than the other. For now, though, exactly what's happening isn't known.

We still haven't answered the initial question: how do the snails know if it's a trail left by a friend or by potential food? The answer lies in the slime. Dr. Harrington showed this clearly in a lab experiment. First she collected wolf snail slime and slathered it onto one of their normal prey, a snail called a helix snail. "If we covered the helix snail with the wolf snail slime," she says, "the wolf snails ignored it as if it were their own species. When we did the opposite experiment – if we took a wolf snail and rubbed a prey snail's slime over it – then the snail that was wearing the prey slime was a goner. The predatory snail, as soon as it tasted the slime, would attack. It didn't matter that what it was attacking was actually a wolf snail."

Considering how small its brain is, this snail is quite an advanced creature. Dr. Harrington says, "Their brains are a billion times simpler than mine, but somehow, they know their right from their left. And maybe I'm not as smart as the average person, but I was in college before I could reliably tell my right from my left. These little guys don't even have a backbone and they know their right from their left. That's one thing that impresses me."

They might be slow-moving, but they're nothing if not quick-witted.

How a Bat Holds its Licker

The wonder of nature is that most creatures are perfectly adapted to their environment, even if their adaptations sometimes take a bizarre and extreme form. Take, for instance, the nectar bat

(*Anoura fistulata*). Its extreme adaptation is a tongue that would make Gene Simmons and the members of KISS envious. Dr. Nathan Muchhala, now a researcher at the University of Toronto, discovered this bat's unusual lingual licker.

This bat is a native of Ecuador, where it lives on the nectar of various flowers. It's not particularly large, with a body length of about 6 centimetres (less than 2.5 inches), although it does have a larger snout than the typical bat. The size of the snout isn't a surprise given the length of the tongue. It's a whopping 8.5 centimetres long (3.3 inches), or roughly one and a half times the length of the bat's body. On an adult human male, that would translate to a tongue over 2 metres (or yards) long! If you're wondering how on earth a tongue like that fits into the bat's mouth, the answer is that it doesn't. The bats have come up with a unique way of storing the tongue. In mammals, including other bats, the tongue is attached to the back of the throat. But not in these bats. Their tongue goes all the way down the throat and into the chest. It's actually attached down between the rib cage and the heart.

If that's not weird enough, even the appearance of the tongue is odd. When it's stored away, it's quite fat and relatively short. Then, when the bat sticks out its tongue, it contracts the muscles, making the tongue longer and thinner. Dr. Muchhala says, "The tongue is huge. It's a skinny, extendable organ that's pink and it has kind of a neat adaptation. It has bristles on the end that form kind of a mop, to mop up the nectar."

Here's what you'd see if you were to watch one of these bats feeding. First, it would approach a flower called *Centropogon nigricans*. These flowers have a long, tube-like structure, at the base of which is the nectar. The bat hovers in front of the flower (much like a hummingbird), then it flicks its tongue out, right down to the bottom of the tube, and scoops up some nectar. This happens really quickly – so quickly that the bat will repeat this three or four times in half a second. Not too shabby when, as Dr. Muchhala points out, "This would be like a cat being able to drink milk from its dish at a distance of two feet [sixty centimetres] away."

This is a very specialized type of adaptation. Over time, the plant's flower has become longer and longer, probably in response to an elongating tongue in the bat. This might seem counterproductive; after all, the flower is totally dependant on the bat for fertilization, while this bat can survive quite well with the nectar of other flowers. So why would the flower make feeding difficult for the bat? It's probably because when the bat sticks its head into the flower to get the nectar, it gets dusted with pollen. Then, much like a bee, the bat will transfer that pollen to the next flower it visits. If the flowers are too short, then only the tongue goes in, the pollen never gets on the bat's head, and the next flower doesn't get fertilized. The longer the tongue, then the longer the flower has to be. Over enough time, this incredible record-breaker has evolved. No other mammal has a longer tongue relative to its body size.

In case you're wondering how Dr. Muchhala measured the length of the tongue when the bat uses it so rapidly, it was actually pretty simple and didn't harm the bat. He just had to train his bats to drink from a straw. Once they'd do that, it was simply a case of putting the nectar farther and farther away from the straw's opening and measuring how far the bats could reach. He already knew they'd be able to reach pretty far – all nectar bat species can. But when he found that the tongue was half as long again as the body, he was, in his own words, "blown away."

It's a good job none of these bats was upset by the experiment. Can you imagine the raspberry they could blow?

Its Bite is Worse than its Bite

In the terrifying science fiction movie *Alien*, a frightening monster invades a cargo ship and proceeds to eat the crew. One of the scarier aspects of the alien is that when it opens its huge jaws, a second, smaller, sharp-toothed jaw shoots out. It's one of the most frightening images of a creature ever invented, and it could only have come from a nightmarish imagination. Except that nature has, as usual, anticipated our nightmares. The movie alien is just a pale imitation of one of evolution's innovations. Dr. Rita Mehta, a biologist at the University of California, Davis, has discovered that the fearsome moray eel (*Muraena retifera*) has a similar double jaw.

The head of the eel is long and skinny, and the mouth is very wide. At the back of the throat, near where it closes off to go down to the stomach, is where this second set of jaws is found. This set looks pretty much like the first set of jaws, only smaller. It has teeth, and the jaws can open and close and grab on to prey. They're called pharyngeal jaws, and in fact, most fish have some variation on them. In most fish, though, the pharyngeal jaws are simple plates that they use to crush and grind prey. Only the moray eel has developed them into jaws that look like a small version of the main set.

You might shudder to learn that these jaws are extendible – that is, the eel can bring them forward into the cavity of the mouth. When the eel catches its prey, the pharyngeal jaws move forward, grab the food, and pull it back and down the throat into the esophagus. In a sense, it's a mechanical way of swallowing food.

It turns out that swallowing is what the second jaw is for. When Dr. Mehta brought some moray eels into the lab to study, she noticed that they fed only by biting. "They have a very reduced capacity to suction feed," she says. "And this is really interesting in light of what all other fishes do. The majority of fishes use suction to capture their prey. And they create suction by rapidly expanding their mouth cavity." With their long, thin heads and wide mouths, moray eels have lost the ability to create this kind of suction. That's not a huge problem for grabbing food that's passing by in the environment, but it is an issue if you want to get it down your throat. Imagine your own ability to eat if you couldn't physically swallow. So, the second set of jaws replaces the suction and gets the food all the way down to where the throat muscles can take over.

It's a novel system of feeding that escaped the attention of researchers for a long time. Researchers studying these eels in the 1960s saw these jaws, but no one thought to take a look at what they were doing. No one thought they were used for swallowing. Dr. Mehta herself said, "No way. That's amazing!" when she saw them for the first time.

Talk about chewing your food properly!

Whales Lunge for Lunch

Baleen whales seem to have it easy. After all, when they feed, all they have to do is glide through the water with their mouths open. The krill they munch on float in the water like a cloud – the whales don't need to hunt the krill down, or fight with them. All they have to do is open their mouths and swallow. But looks can be deceiving. Most baleen whales, which include the giant blue whale and the more familiar humpback, actually have to work quite hard to get their food. Jeremy Goldbogen, a Ph.D. candidate in zoology at the University of British Columbia, is the one who discovered just how much the whales work when they're hungry.

Just to be clear, though, this doesn't necessarily apply to all baleen whales. What Mr. Goldbogen has been studying is a group of baleen whales, known as the rorquals, which are made up of the fin whales, blue whales, minkies, and humpbacks. Physically, they're different from the other baleen whales. "They're distinguished by a very big mouth. They have ventral pleats that extend from the snout all the way to the umbilicus," Mr. Goldbogen says. (Picture a sieve of overlapping vertical bars attached to the upper jaw.) "Underneath that, at the floor of the mouth, they have this very floppy, extensible, specialized blubber layer that we call the ventral groove blubber."

These animals really do have big mouths. Their jaws are up to one quarter of their body length, tail fin included. When they open their mouths all the way, which involves dropping their bottom jaw ninety degrees (i.e., straight down), those ventral pleats unfold and the mouth expands to four times the size it was when it was closed. It's as if they've opened a parachute.

This parachute mouth has the effect of slowing the whale down almost to a full stop when it's opened. This is the same drag effect used to slow down the space shuttle when it lands, or those speed-record-setting cars that occasionally race through the desert. But if their parachutes are released when the vehicle is still, they'll just fall to the ground. The same is true for the whale. If the whale isn't swimming, its mouth won't open properly. In order for its mouth to work, the whale needs to be moving at about three metres per second (seven miles or eleven kilometres per hour), which is a fair clip underwater.

This has led to a behaviour called ram feeding, or lunge feeding. The whale gets up a head of steam, approaches a group of krill, opens its mouth, and takes in a huge volume of water and krill. Then it closes its mouth, which is now full of water and krill. The water flows back out through the baleen plates, and the krill stay in the mouth. Mr. Goldbogen says, "We've estimated that [the whale catches] about twelve kilograms [twenty-six pounds] of krill per gulp, or per lunge." Not a bad-sized mouthful.

This method of hunting – racing in, opening the mouth, coming to a rapid stop, and then swallowing – uses a lot of energy. For Mr. Goldbogen, the question was how much. Researchers had noticed a paradox. Normally, the bigger the mammal, the more efficient its metabolism becomes. For whales, that means the bigger they are, the longer they should be able to dive without running out of air. But no one seems to have told whales that. While large sperm whales can make very long dives, the blue whale, also large but one of these lunge feeders, seems to make only relatively short dives – "relatively" because the dives can still last twenty minutes, which is long for a human. But compared with a sperm whale, which can stay underwater for more than an hour at a time, the blue whale's dives don't last long. The same is true for humpbacks; they also spend only about twenty minutes at a time under the surface.

Even beyond dive times, there's more evidence that these whales are really pushing themselves when they go hunting. Anyone who's ever exercised hard will understand this next part. Whales that have been lunge feeding spend longer at the surface breathing between dives. You can see this in humpbacks. If they've been down singing, their surface breaks are short. But when they've been hunting, they have to spend three times as long at the surface before they're ready to go down again. This isn't to suggest that whales are actually panting at the surface, but they're definitely taking more breaths between dives when they're lunging. And the more they lunge, the shorter their dives.

So much for the idea that floating food is easier to catch. When you have a big mouth, you need to fill it with a lot of food, just so you have the energy to close it again.

Killer Whale Sonar

If you've seen a killer whale (*Orcinus orca*), you'll know they're impressive predators. The killer whales that live off the coast of British Columbia and Washington State are such good hunters that they can afford to be picky about their prey. They don't go for just any seafood platter. No, they go for the good stuff. In the Pacific, that's the chinook salmon.

Scientists have been wondering for a while just how killer whales manage to select only the chinook. They make up only about 15 per cent of the salmon swimming in these waters, so picking them out of a school of more abundant and similar-looking coho and sockeye salmon has to be a daunting task. Dr. John Horne, a marine mammal specialist at the University of Washington, thinks he's figured out the key to this finicky foraging. He says the success lies in the sound approach taken by the whales.

The killer whale's preference for chinook salmon is quite remarkable. A couple of decades of Canadian research by Dr. John Ford, at the federal Department of Fisheries and Oceans in Nanaimo, B.C., shows that no matter what fish are out there, the killer whales will always pick the chinook given the choice. Considering how much easier it would be just to scoop up any old fish, they're working hard to mark their preference. Also, telling salmon species apart isn't easy. If specimens of the three main species were placed directly in front of you, police lineup style, you could probably point out the differences. But if they were shown to you one at a time, you'd have a hard time working out which was which. The differences are subtle. Now try this underwater, in the dark, and you get an idea of what the whales are up against.

The whales do have some tools at their disposal that we don't. Killer whales are actually part of the dolphin group of marine mammals. And like their cousins the bottlenose dolphins, these black-and-white behemoths have a sonar system. There are two types of calls the whales use: communication calls and echo clicks. The first is fairly straightforward – just shout-outs to the neighbours. The echo click is more complicated. Dr. Horne says, "They send out a short burst of sound, of energy, through the melon [the dome] and upper jaw of the head, and listen for the return or the reflection of that energy off a fish or another potential prey item. When it comes back and is received, it's transmitted through the lower jaw up into the brain." This is the sonar system, which shares a lot of characteristics with the sonar that bats use to echolocate their prey.

Simply bouncing signals off a fish might tell you there's an animal there, but telling the species apart requires a great deal of subtle interpretation of the signal. For that, Dr. Horne did experiments in the lab.

His first step was to build a lab sonar system. Then he got samples of the three species of fish from a local fish farm. He placed the fish in front of the sonar at different angles, which required, as Dr. Horne says, "anaesthetizing them and tethering them to a monofilament line." Then he studied how the fish reflected back the sound. It's like taking a flashlight, shining it off an object, and seeing what the reflection looks like from different angles.

When Dr. Horne listened to the clicks that came back, they were hard to tell apart. But when he ran the same signals through his computer and looked at the wavelength pattern, sure enough, the three fish did look slightly different. Presumably, it's that difference the whales are picking up on.

What's probably making the difference to the wavelengths is a structure in the fish called the swim bladder. The role of the swim bladder is to hold gas, to keep the fish buoyant. It's a very different density from both the water around the fish and its flesh, and this makes it extremely reflective to the sonar. Chinooks have a smaller swim bladder than either of the other two fish, and on top of that, their swim bladder sits at a different angle within the fish. A change in this angle will also change how the signal is reflected.

Somehow, and it's not clear yet exactly how, the whales are sensing this very small set of differences in the echoes from the three types of fish, and using the information to pick out the chinook salmon. And that's definitely to their advantage. Of the three species, the chinook has the highest fat levels. This not only makes the chinook the most energy-rich food option, it improves its flavour.

Who says whales don't have good taste? In this case, the killer whales of British Columbia and Washington State are true gourmets.

SEA LION DIVING

One growing problem in the North Pacific is the fate of the stellar sea lion (*Eumetopias jubatus*). The population numbers have crashed from Alaska across to the Aleutian Islands. Part of the problem may be predation by killer whales, but researchers think that a number of different factors – climate change, changing ocean conditions, changing fisheries, the kinds of food available, and just living in the north – are probably all acting together in this case. To untangle all these factors, first we need to know more about the sea lions themselves – to understand the basics of their biology, for example, how much energy they use to roam the sea and hunt their food. Dr. Andrew Trites, director of the Marine Mammal Research Unit at the University of British Columbia, has taken a unique approach to his research, using sea lions themselves as part of his team.

He's studying the sea lions' diving behaviour, to see if his findings fit with the hypothesis that the decline is at least partly due to overfishing. The more fish we take, the deeper the sea lions have to dive to find their food. If this is causing them to use more energy, then they'll need to get more food, and a vicious cycle develops. But before we can place the blame firmly at the feet of the fisheries, we need to know how much energy the sea lions are actually using, and whether there's an optimum depth for them to capture their prey.

Studying wild sea lions presents a problem. "One of the challenges in studying animals in the wild," says Dr. Trites, "is that you can't get access to them every day of the year." Not only that, but wild animals are lousy at following instructions. Dr. Trites has come up with a creative solution. "We've been working with stellar sea lions at the Vancouver Aquarium," he says. But the pools they live in are only about 4 metres (13 feet) deep, and sea lions can easily dive up to 100 metres (328 feet), sometimes going as far down as 300 metres (almost 1,000 feet). "So right now," he says, "we have what we call their open water research project, where we have five stellar sea lions trained to swim freely in the ocean with divers. We take them up to Indian Arm, in North Vancouver, where we have some extremely deep water. The animals are now trained to do diving studies with us, so we can learn how they find prey, and the trade-offs that they make to do it." Yes, that's right, he's recruited the sea lions to work for their lunch. The sea lions don't seem to mind. "They're completely at ease," Dr. Trites says, "carrying devices on their backs, moving into small chambers. They work 100 per cent freely with us." And they work for fish!

Dr. Trites and his team have even built a special craft, called the Stellar Shuttle, to take the sea lions from the Vancouver

Aquarium, where they live, to the research site. "It's a front-loading craft," he says. "The animals jump on board, and we drive them out to the site. And there we have a diving platform." Here's where the training of the seals pays off. Dr. Trites's team has a series of tubes that screw together to make different lengths of piping. The seals are trained to swim down to the end of the tube and hold on until they get a signal to let go – either a light turning off, or some fish pieces being released from the bottom of the tube. Then they swim back up.

To determine how much energy the seal is using, Dr. Trites measures the strength of the oxygen in their breath. Ever seen one of those bicycle set-ups where the rider has a mask attached so that the researchers can measure how much oxygen the rider is using? Well, that's not terribly practical gear for a sea lion. First of all, they wouldn't put up with wearing the mask. Second, while they're underwater, they're holding their breath, so it's not as though you'd have much to measure. Dr. Trites has devised another clever solution. He's floats domes on the surface of the water. The sea lions are trained to pop up from their dive under these domes, which collect the air they exhale. By measuring the changes in the oxygen content under the dome while the sea lion is there, the researchers can figure out what's going on with the animal's breathing, and even how much oxygen they used during the dive.

The results Dr. Trites has collected so far are surprising. "We saw a number of different things. One of the puzzling things was, for example, we found that it was cheaper for a sea lion to dive than just to stay on the surface. The animals are actually conserving energy as they dive, by up to about 43 per cent. They do this by slowing down their heart rates and reducing the blood flow. But the other puzzling thing to us was we noticed that as the animals were diving, the shallow dives seem to cost more energy than the deeper dives. At five metres [sixteen and a half feet], the animals seemed to be using a lot more energy than they were down at thirty metres [almost one hundred feet]."

The reason for this lies in the fact that sea lions are pretty buoyant animals. That means it takes them a lot of energy to get from the surface to under the water. When diving to a shallow depth, they're continually fighting to avoid being pushed back up to the surface again. But when they get deep enough, the pressure means their lungs begin to collapse, and when the lungs collapse, they lose buoyancy and start to drop rapidly. "Like a stone," says Dr. Trites. Now it's easy for them to stay down; they're essentially free-falling. That also means it's not going to take much energy to get back to the surface, since they just need to swim up to where their lungs expand and they start to float again. Diving deep is actually so efficient for these animals that they can reduce their daily calorie requirement from eighteen kilograms (almost forty pounds) of food a day to eight kilograms (less than eighteen pounds) if they do all their dives to around three hundred metres (a thousand feet) below the surface, rather than staying in the top ten metres (roughly thirty feet).

So, scratch the hypothesis that being forced to dive deep might be leading to problems for the sea lions. In fact, it's their preference.

There's one more interesting discovery Dr. Trites has made with his trained sea lions and their breathing domes. It had been noticed before that when a sea lion finds a patch of food in the wild, it will make several dives in a row to catch the food. When the research team simulated this with their sea lions, they found that the more dives the sea lion took, the more energy each successive dive took. As Dr. Trites says, "They were not refilling their oxygen tanks between each dive, if you like, until the very end."

A case of catch the fish while they're there, worry about catching your breath later. Of course, this is a limited strategy. Eventually the quick breaths at the surface would mean the carbon dioxide levels in the blood would be too high for the sea lions to keep going. But for the amount of time the fish are there, it works quite well.

We wonder what other animals could be recruited for studies like this in the future. Just so long as it doesn't put research assistants out of business.

DOLPHIN HERDERS

If you've ever watched a football game, you know how complicated it can be. Each team has the same goal, but each player has a very specific role and detailed instructions for each type of play. Teamwork is something humans do very well, which may be a leftover from our early days of hunting co-operatively on the African plains. But other groups of animals aren't generally as well coordinated. They might hunt in groups, but even so, it's usually a case of everyone rushing at the prey and seeing who gets to knock it down. That is, unless you're talking about dolphins. Their hunting techniques may be as nuanced and complex as any football drill. This behaviour was discovered by Stephanie Gazda while she was a Ph.D. student at the University of Massachusetts.

This coordinated behaviour was first seen among a population of one hundred resident dolphins that live near Cedar Key in Florida. Ms. Gazda focused in on two groups. One was a group led by a dolphin named Tall Fin, the other was a group led by a dolphin named Point. Tall Fin's group usually contained three dolphins; Point's sometimes had six. But both of them had similar, specialized hunting patterns, called a driving behaviour.

This driving behaviour would always start the same way. Tail

Fin, or Point, would start swimming in a circle. Ms. Gazda says, "They use a very distinctive method of swimming. It's like they're very actively pumping their tails out into the water, and they go into circles, and then the non-driving dolphins line up as a barrier." While the driving dolphin is turning circles, it is trapping the fish and driving them towards the dolphins that form the barrier. The fish start panicking and jumping out of the water. "Then," says Ms. Gazda, "the dolphins are able to grab fish flying out of the water. They're actually partially out of the water as well, and they're just grabbing fish." Easier than shooting fish in a barrel.

What makes this whole behaviour unusual is that the roles seem to be very tightly fixed. It's always Tail Fin, or Point, who drives the fish towards the other dolphins. The others always form the wall. While other species hunt in groups, this kind of specialization is found only in dolphins and humans. And, on some levels, it doesn't make sense. Tail Fin and Point are working a lot harder than all the other dolphins in their groups to isolate the fish and drive them to the barrier. Ms. Gazda thinks this role specialization may slightly improve the hunt, but she doesn't have solid data to back it up.

It's also not the only hunting method these dolphins use. They sometimes hunt alone, either randomly foraging or following their own specific patterns. "But," says, Ms. Gazda, "the fact that we see the driving behaviour so frequently, and we would see it for long periods of time during the day, means it has to be efficient; otherwise, there's no reason for them to do it."

At this point, we don't know how widespread this behaviour is. The human residents of the Key have seen the activity for years and didn't think it was special. It's possible other groups of dolphins have similar behaviours. We do think dolphins are extremely intelligent. If they're capable of this kind of coordination, that's another sign they're a lot more like us than we might think.

And, just maybe, the Miami Dolphins are getting their playbook from their namesakes.

Guano-Eating Salamanders

Living in a cave isn't easy. But the grotto salamander (*Eurycea spelaeus*) seems to have most of the problems of cave life covered. It's dark in there, so the salamander uses senses other than sight to get around. In fact, this species is blind. The cave is also cold and wet, which is heaven for a salamander. And then there's the issue of meals. There's not much food in a cave, and what there is is of very limited selection. This is a serious problem for the salamander. But according to Dr. Dante Fenolio, an amphibian specialist at the Atlanta Botanical Garden in Georgia, they've devised a solution to this problem that's as creative as it is disgusting.

The caves where the grotto salamander is found are in the Ozarks. As Dr. Fenolio describes the area, "This part of the Ozarks is a big piece of Swiss cheese, and in northeastern Oklahoma they're riddled with caves and networks of subterranean passageways. It's a really exotic habitat to be inside of. You're wading through a river up to your chest. In my case, I'm a little shorter so maybe up to my neck. And you've got stalactites and the stalagmites and this amazing river." The wildlife in these caves consists of salamanders, bats, and a blind, colourless species of crayfish. Neither the crayfish nor the bat makes a good meal for an amphibian, and beyond them, there's not much for the salamanders to snack on.

Dr. Fenolio wanted to find out what they were eating. Normally, salamanders are carnivorous predators, living on small bugs and sometimes small fish. But there weren't many, if any, of those prey in these caves. One day, while collecting salamanders in plastic bags to study them, his colleague, Jim Stoud, saw something strange. "He noticed," says Dr. Fenolio, "pieces of guano in some of these bags, and he was sure that the bags were clear when we first put the salamanders in them. And then, finally, we caught one of these salamanders that was actually chewing on a piece of guano."

Yes, that's right. The salamanders are eating guano, which, for the uninitiated, is bat poop. These salamanders are swimming up to the surface of the water, after the bats come in to roost, and munching the fresh guano that the bats poop while they sleep. This is highly unusual behaviour for something that's normally a carnivore. It appears that the salamander has had to adapt to an omnivorous diet.

It seems to be a smart choice for the salamanders, though. When Dr. Fenolio analyzed the bat guano, he got a surprise. "We realized very quickly," he says, "that the nutritional value of guano was very high, particularly when you put it into comparison with some of the invertebrate prey items that these salamanders eat. We compared it to a hamburger, and if you look at the vitamins and minerals that are still left in it, if you look at the caloric content, etc., it actually exceeds the hamburger." That's right: Looking for a nutrition fix? Then try bat guano.

It does make sense. Bats have developed a very short digestive tract. That means that much of the food they eat doesn't get digested particularly well on the way through. While this might seem like a waste of effort, catching all those bugs every night, in fact, the bats can't afford to hang on to their food too long. If they do, it weighs them down, and they can't fly properly. Their guano, then, is full of half-digested insects, which is free nutrition available for the salamanders.

The guano likely smells of the half-digested insects. That's probably what attracted the salamanders in the first place, and, over time, these amphibians have adapted to this novel food source.

It takes them into a rarefied group, the coprophages, organisms that eat dung. Some tortoises eat hyena feces, because the hyena have eaten a lot of bones so their poop contains high levels of calcium, which the tortoises need. And then there are the animals, like rabbits, that eat certain types of their own droppings. But it's not a particularly common habit.

For the salamander and the bat, though, it's a great combination of food source and housekeeping.

The Hummingbird's Stopwatch

So many flowers, so little time. How true that is for a hungry, hovering hummingbird, poking through the petals to collect nectar. But as there are lots of flowers out there, how do the tiny birds tell if they're visiting a flower they've already emptied of nectar? After all, there's no point wasting time opening an empty cupboard. Dr. Andrew Hurly, a biologist at the University of Lethbridge, Alberta, has been studying time management in hummingbirds and thinks he knows the answer.

Rufous hummingbirds (*Selasphorus rufus*) are found across western North America, particularly in forested and mountainous areas. They're about eight centimetres (three inches) long, weighing in at about three grams (a tenth of an ounce). They look like any other hummingbird, with rapidly flapping wings and a long beak for getting down into flowers. Their name comes from the colour of the males – they're brown, specifically rufous brown, which is a rusty brown.

The issue hummingbirds face is a simple one. Their wings beat about seventy times per second, and their hearts beat a thousand times a minute, which makes for one huge metabolic rate. This takes a lot of energy to support, and the birds can't afford to waste any effort. At the same time, individual flowers contain only a tiny drop of nectar. That means the birds need to visit hundreds of flowers every day to get enough energy. Every empty flower the hummingbird visits is a waste of energy. But flowers do eventually replenish their supply of nectar, so if a hummingbird can work out when to come back to the same flower, it can save energy by not having to fly so far afield for food

It was this time management skill, their knowing when to revisit, that Dr. Hurly decided to study. Could hummingbirds identify which flowers they'd been to already? And would they know when it was a good idea to come back?

To test this, Dr. Hurly set up a simple experiment. Within their own wild territory, the hummingbirds were given access to eight fake flowers, each a different colour, so that the birds could tell them apart. Four of the flowers would get a hit of nectar ten minutes after the bird emptied them, while the other four were replenished only after twenty minutes. The results were impressive, Dr. Hurly says. "What we found was that even within the first day, the bird started to treat the ten-minute and the twenty-minute flowers differently. On average, they would visit the ten-minute flowers every ten minutes and the twenty-minute flowers every twenty minutes."

So, not only were the birds able to figure out that the flowers were refilling, they could distinguish, and remember, the different time frames for the two different conditions. This goes far beyond the capacity of a stereotypical bird brain, which in the case of these hummingbirds is the size of a dried pea. And it didn't take them long to learn. "They seemed to start figuring it out within the first few hours, which we found astonishing," says Dr. Hurly. They did take longer to work out the timing of the twenty-minute flowers than the ten-minute flowers, but within a couple of days, they'd stopped making any mistakes.

And remember, each of the flowers in the territory was a different colour. "The analogy," says Dr. Hurly, "is as if the hummingbird had, in its brain, at least eight separate stopwatches, and every time it emptied the nectar from the flower, it restarted the stopwatch for that flower. And most of the time, it didn't come back to that flower until that watch had timed out in either ten minutes or twenty minutes." Most of us would have trouble juggling that many different variables, but the birds seem to have no problem. While the experiment didn't test it, the birds are probably making the same calculations for all the flowers they visit in a day, which is well beyond what most humans can manage.

Add to this the fact that the birds also have a great sense of place. Other research has shown that once they know where a given feeder or flower is, they will note if it gets moved. The one variable left is whether they can tell the difference between different types of food, and pick out only those that are the best. All this, while flying from plant to plant, and without a stopwatch. Forget busy as a bee – it should be busy as a hummingbird.

How to Avoid Becoming a Meal

While some are the hunters, others are the hunted. Here are some examples of how the hunted can fight back, from using makeshift claws to donning elaborate disguises.

Wolverine Frogs

Dr. David Blackburn, an evolutionary biologist now at the University of Kansas, has spent several field seasons in Africa studying rare and new frog species, so he's familiar with the ways of amphibians. Which is probably why he got such a surprise the first time he picked up a hairy frog (*Trichobatrachus robustus*). The frog inflicted some deep and painful scratches. Once he was over the shock, Dr. Blackburn started to wonder how the heck that had happened. Unlike cats, dogs, humans, and many other animals, frogs don't have proper claws or nails. And yet, this species seemed to have them. Dr. Blackburn's subsequent studies of this frog have led him to some amazing, and slightly disturbing, conclusions.

He remembers the day he was scratched: "At the time, I was doing field work in Cameroon, and I was in the mountains in Southwest Province. And we were just, as part of our normal field work, going around collecting animals, and I picked up a large female hairy frog and suddenly got a nasty scratch on my hand. And it was because the female had just started kicking her feet wildly when I picked her up.

"Usually, when you're out catching frogs, it's a gentleman's-type sport, you know? You don't have to worry about getting scratched, or bitten, or harmed in any way." Dr. Blackburn thought that maybe he'd picked up something else along with the frog, but no, that was all he was handling. Which led him to a closer examination and an understanding of what had happened.

First of all, there isn't really a claw. The hind feet of these frogs (which is what they use to scratch) is just like any other froggy foot out there. "Instead," says Dr. Blackburn, "the last bone of the finger is actually able to pierce through its skin and come out. Now, normally tetrapods – birds, mammals, reptiles – do a pretty good job of keeping their bones in their fingers. This is not a common occurrence." Not a common occurrence, indeed. Think about it for a moment. This frog's bones can be pushed through the skin and used as claws. If you've seen the *X-Men* movies, or read the comic books, this will seem familiar, since it's what the character Wolverine does. However, in these frogs, it's the skin on the underside of the foot that's pierced, not the top, as it is with the comic book hero.

Even the piercing of the skin itself is remarkable. The bone looks like a claw. It's slightly curved and comes to a very strong point that's reinforced at the tip. When the frog flexes its foot, this bone is released from a structure that normally holds it in place, and it shoots through the intact skin.

Dr. Blackburn has seen the bone claws retract back into the feet, but it isn't clear whether this is an active or passive process. Cats, for instance, have passive retraction – there's no muscle pulling the claws back in. It's likely the process is passive in the frog, too, but no one yet knows for sure.

It's also not known how destructive this behaviour is for the frogs. Dr. Blackburn has never seen any blood from the frog after they've pierced their skin with the claw, just his own. And what happens after they've retracted them isn't known either. This is partly because it's hard to follow these frogs in the field, and no one's worked with them in the lab to experiment on what happens to the feet. Given that we're dealing with amphibians, it's possible that they can regenerate the skin and any other damaged tissue, but someone needs to do the experiment to find out for sure. For now, the work is being done mostly with museum specimens rather than live frogs, so the anatomy is known, but the physiology is a mystery.

One thing that is reasonably clear is that this is a last-ditch technique for the frog. "These wounds really look traumatic," says Dr. Blackburn. "The claw busts right out and tears a nice little gash in the end of the tip." Which suggests that it would leave the frog susceptible to infections, and wouldn't be the kind of behaviour you'd expect to see very often.

The next time you're in Africa, you might want to avoid picking up a frog. It might give you a nasty scratch.

BOMBARDIER BEETLES

No matter how creative and clever we get, nature can always put us in our place. We can boast about our high-powered computers and brilliant engineers and remarkable manufacturing techniques to make our technological marvels, but nature has had a couple of hundred million years of evolution to get her engineering triumphs right. We may think we're hot stuff, but we can still learn a trick or two from her. For instance, take the defensive armament of the bombardier beetle (family Carabidae).

This little insect can squirt its enemies with a high-pressured jet of boiling, toxic liquid from a flexible nozzle at the tail end of its abdomen: a feat we can only wonder at. We'd like to understand a little more, however, and maybe steal a few of the beetle's trade secrets for use in such things as jet engines. That's what Professor Andy McIntosh, an engineer from the University of Leeds in England, is hoping to do. He's taking the secrets of the bombardier beetle and applying them to the business of keeping planes in the air.

The bombardier beetle's high-powered squirt system is an astonishing piece of engineering. First of all, the beetle is able to direct its nozzle, which comes out of its backside and is about two centimetres (three quarters of an inch) long, in any direction. "Say an ant attacks one of its legs," says Professor McIntosh. "It just turns this nozzle and squirts hot fluid, mainly water. It's about 100 degrees Centigrade, and it comes out as a combination of steam and hot water, plus a noxious mixture of hydroquinone and hydrogen peroxide. It's a most amazing sight when it's done."

The mechanism behind this is ingenious. Inside the back end of the beetle are a series of chambers, which together look roughly heart-shaped. In one lobe of the heart, there's hydroquinone, and in another is hydrogen peroxide. These are mixed in a third compartment, where they start to react. The reaction, though, is slow, so the beetle adds in something called catalase, which is a molecule that catalyses, or speeds up, the reaction. This creates an explosion, and the high-pressure liquid (which, thanks to the chemical reaction, is now mostly boiling water mixed with steam) shoots out its posterior. This then creates a low-pressure area in the compartment that's been emptied, which draws in more of the hydroquinone and hydrogen peroxide. Add a touch more catalase and the whole cycle starts over.

The beetle will do this twenty to thirty times in a row, expelling this very fast stream of boiling liquid. There's a quiet hissing sound as the liquid flies out in rapid pulses. Oddly enough, humans have devised something very similar. The V1 rocket, used by the Germans in the Second World War, was a pulse engine and had the same basic principles of operation. However, the beetles do it faster, and their jet is a lot more efficient than the human version.

Engineers do want to use this technique to make more efficient jet engines, but probably not in the way you would think. Unfortunately, it's possible for jet engines to go out while in flight. (If you fly a lot, don't think about this for too long.) Restarting them isn't easy. You need to run a huge electrical current through an electrode. This splits the air around the electrode into ions, and these ions are then shot into the engine, where they can relight the kerosene fuel that's there. The problem is, it doesn't always work, and it has a very limited range.

That's where the beetle comes in. It all comes back to that chamber where the two chemicals mix. "The little chamber is only one millimetre [0.039 inches] long," says Dr. McIntosh, "and yet it's able to send a jet about two centimetres [three quarters of an inch], which, for the beetle, is quite a long distance. Therefore, we reckoned on a heart-shaped combustion chamber, plus the nozzle. When we actually scale it up to about two or three centimetres [around three quarters of an inch, an inch and a half], it might be able to eject its mass when you have an explosion inside. It may be that mass will go about twenty centimetres [almost eight inches]." That would be far enough to relight a jet engine.

A similar idea is using this technique to ignite airbags. In case you didn't know, an airbag goes off in your car when a gas generator inside the bag is ignited. Dr. McIntosh thinks the speed with which the bombardier beetle's jet fires could provide a model for ignition systems everywhere, particularly in airbags.

Biomimetics, the science of copying the natural world, is full of surprises. And while this work isn't done yet, who knows what else the bombardier beetle can teach us? Just don't stand behind one and scare it – that stuff burns.

TOADFISH PEE

It's called "waste" for a reason – the stuff we don't need, whether it's in the garage or in our bodies. And it all needs to be disposed of in an appropriate manner. All animals have to deal with waste products. For most, it's just a case of excreting what you don't need and moving on, but that's not so for the toadfish (*Opsanus beta*). You could argue that the toadfish has become a master of recycling; at least, that's what Dr. Patrick Walsh, a biologist from the University of Ottawa, would tell you. He's discovered that the toadfish puts its waste to a disarming use.

Toadfish are not the most attractive fish out there. Dr. Walsh says, "I like to think of them as having a face only a mother toadfish would love. They're kind of flattened from top to bottom, they have a big, wide mouth, mottled brown, blotchy skin that blends in with their environment, and they're pretty nasty little cusses as well. If you stick your finger in their tank, they'll go after it." If you want to avoid having your fingers bitten, be careful when you're visiting the southeast coast of Florida or the Gulf of Mexico, where these creatures are found.

To understand what makes the pee of a toadfish special, we need to make a quick diversion to the chemistry of urination. There are two possible chemicals that animals use to excrete waste products: one is ammonia, the other is urea. Ammonia is pretty toxic; urea is much milder. Biologically, it takes energy to turn ammonia into urea. So for creatures that don't have to worry about running into their own waste, ammonia is the way to go. That's what most fish use. They don't exactly pee; rather, they excrete their liquid waste across their gills. But they excrete it as ammonia, because the water will wash it away from their gills and keep them healthy.

On land, however, the situation isn't as simple. Most creatures store their urine in their bodies and only eventually pee it out. For that reason, it's worth changing it into a less-toxic urea. Also, when you change ammonia into urea, it frees up some of the water, which means you're going to be less dehydrated if you pee urea rather than ammonia, since you get to keep more of your own liquid.

Toadfish fall somewhere in between. They produce a combination of both urea and ammonia. This seems very rare in the fish world. Lots of fish produce urea when they're in the egg, presumably to prevent their environment from becoming toxic. But by the time they are adults, they've switched to producing ammonia. The question then becomes why a toadfish would waste the energy on making urea.

Strangely, the answer may have something to do with their voices. Toadfish are loud creatures. "They have a large sonic muscle attached to their swim bladder," says Dr. Walsh, "that they use for aggressive grunts at each other, as well as a very loud and interesting call that the males use during mating season, called the boat whistle." How to describe the boat whistle? Think of a high-pitched, very fast raspberry sound, in two blasts like the whistle on a boat, then you'd be close. It's an odd sound coming from a toadfish, to say the least.

Being loud has its disadvantages. It makes it easier for you to find mates, but it also makes it easier for predators to find you. And here's where the pee comes in. It turns out that the urea-ammonia mix may cloak their location from predators. So, a predator hears a toadfish, swims close, and then, if it were any other species, follows an ammonia trail into the burrow and grabs the prey. The urea, while expensive to make, confuses the predator.

Dr. Walsh tried this idea in the lab. He set up, as he calls them, "toadfish condos," with clay model toadfish inside. Then he played toadfish calls and released one of their predators, the grey snapper. Adding ammonia to the condos drew in the snappers really well, but if there was a fifty-fifty ratio of ammonia to urea, the number of snapper attacks dropped by half.

If some urea is good, why not go whole hog and produce only urea? Dr. Walsh thinks this is an example of efficiency in biology. The toadfish produces enough to protect itself; more would just be a waste.

Talking of waste, there's an interesting implication of this work. It suggests that in the marine environment, predators may be using ammonia, which is a type of nitrogen, as a way to find their prey. But we're dumping all kinds of nitrogen products into the water. Does that mean, beyond the phytoplankton blooms that have been in the news, we also might be confusing predators and upsetting the normal balance?

SQUIRRELS EATING SNAKE SKIN

Throughout history, humans have applied scent to our bodies to disguise an unpleasant odour, or simply because it makes us smell nice. And nature has always been a good source of these aromas, from the rosewater of the Middle Ages to the lavender of the 1900s to the musky perfumes of today. But it's unlikely the habits of the California ground squirrel (*Spermophilus beecheyi*) will ever provide inspiration for a new fragrance. This rodent has a weird approach to perfume: it covers itself with the aroma of its main enemy, the rattlesnake. Dr. Barbara Clucas, an animal behaviour scientist at the University of Washington in Seattle, thinks there's a good reason for these rodents to want to smell like a snake.

Ground squirrels are a relative of the more common tree squirrel, and they look quite similar, with big bushy tails and mostly grey fur. The biggest difference is that they live in burrows underground, rather than up in trees. They are found throughout the western United States and southwestern Canada, although different species are found in different places.

Each of those species has its own predators to worry about. In California, it's the rattlesnake, and how these two creatures interact has been an ongoing area of academic study. For instance, one of the early discoveries about ground squirrels was that the adults are resistant to snake venom. "So," says Dr. Clucas, "they can get bitten by a rattlesnake and they survive." Unfortunately, this doesn't work for juvenile ground squirrels, so they need other tricks to avoid getting bitten in the first place. Which is where the scent story comes in.

This tale starts while Dr. Clucas was working on her Ph.D. at the University of California, Davis, with a random observation by her graduate supervisor. "He was tethering out rattlesnakes and observing the behaviour of the ground squirrels. And he found that after he removed the rattlesnake from the area, the ground squirrels would be very interested in the spot where the rattlesnake had been lying. They would go over and chew on the grass blades, or the ground where the rattlesnake was, and then start licking their body." This is hardly the reaction you'd expect in a predator-prey relationship. With most species, the smell of a predator is repulsive to the prey; for instance, the smell of cat urine will keep mice away (and your neighbours, too). But the ground squirrels were actually attracted to the scent. Not only would they chew the grass where rattlesnakes had been lying, but when Dr. Clucas showed the squirrels the shed skins of rattlesnakes, the rodents would chew up the skin and then lick themselves all over.

To us, this wouldn't be a particularly attractive scent. Dr. Clucas says, "If you've never smelled a rattlesnake, it kind of has a musky odour. Nothing as strong as a skunk, but something similar to that." But to the squirrels, the smell of a snake was oddly attractive. Not to every squirrel, though. While the juveniles and the females were chewing on the skins, the adult males didn't join in. Which gave Dr. Clucas her clue to why this behaviour might be going on. "As well as being visual predators, rattlesnakes also rely a lot on their olfactory sense. And if a rattlesnake comes up to a burrow that smells like it has a rattlesnake in it, or it doesn't smell like a ground squirrel, that might benefit adult females and their pups." In other words, the females and their young were disguising themselves with the scent of the rattlesnake. A roaming rattlesnake would come across a ground squirrel burrow, give it a sniff, and, if it smelled like snake, leave it alone.

In the lab, that's exactly what she found. With the campus collection of rattlesnakes, Dr. Clucas watched to see what the snakes would do if they smelled ground squirrel alone or mixed with rattlesnake scent. Sure enough, the snakes were very intrigued if the smell of squirrel was on its own, but when it was mixed with snake, they lost all interest.

This strategy would not work if the snake could see the squirrels, however. In that case, the female has another trick, literally up her tail. Rattlesnakes are what are known as pit vipers – not because they live in pits, but because they have little pits on their heads that are sensitive to heat. The female ground squirrels can heat up their tails, which fools the snake into thinking that what they're attacking is much bigger and faster than a squirrel, and so maybe not worth going after. This allows the pups to escape capture. But the scent trick isn't going to help at this point. Where it is effective is at night. Even if it confuses a snake for just a moment, it might be long enough for the female to pick up her pups and carry them out of the burrow. That's one advantage of a burrow with multiple exits.

So, if you're wandering through the hills of California, and you're worried about rattlesnakes, maybe there's a lesson to take away from the ground squirrels – just rub snakeskin on your ankles.

ULTRASONIC GOPHERS

As anyone who's spent any time on the prairies can tell you, Richardson's ground squirrels (*Spermophilus richardsonii*), or gophers, are a pretty loud bunch. When they get up on their hind legs and start screaming, their calls can travel far and wide, alerting the rest of the colony to looming dangers. But when David Wilson (a master's student at the University of Manitoba at the time, and now Dr. Wilson, a post-doctoral fellow at the University of Windsor) was studying these alarm calls, he noticed that sometimes the gophers looked as if they were calling out, only there wasn't any sound. It turns out, there's more to a ground squirrel's call than the human ear can pick up.

The ground squirrel frequently calls out in alarm. It's not surprising, really, as their colonies, which can range from a few animals up to hundreds, are prey for all kinds of species. Badgers, coyotes, foxes, hawks, and eagles all like to eat gophers. And every time a ground squirrel thinks it's in danger, it sends out an alarm call. Makes for a loud lifestyle.

This is why Dr. Wilson was so surprised when he first saw a ground squirrel opening its mouth to yell but heard no sound come out. "The first answer to that," he says, "is – oh, the poor squirrel lost his voice. But as we saw this more and more in different populations, we began to wonder, is there maybe something more to it than that? And so, with some special equipment, we realized that there was ultrasound present."

The special equipment was a detector usually used to listen to bats. Bats produce calls in the ultrasound range, pitched much higher than anything the human ear can detect. And it turns out these ground squirrels are doing it as well. Human ears can detect frequencies up to about fifteen kilohertz. These gophers are making calls at about fifty kilohertz. That's even above what a dog can hear!

Not much is known about how the gophers make this noise. There are other rodents, including mice and rats, that make ultrasonic sounds, and we know that what they do is a lot like singing. They force air across their vocal cords, and if it's done just right, the cords vibrate at a very high frequency, beyond what we can hear. Exactly how they can force this, though, isn't clear.

Using ultrasound could be protecting these animals from predators. After all, if we can't hear it, neither can most animals, and that might give the gophers an advantage: they could warn those around them without attracting a predator's attention. And even if the predator can hear in this range, ultrasound doesn't travel very far – say, ten metres (thirty-two feet), compared to the one kilometre (roughly a thousand yards) that a verbal call travels. So this could be a way of warning the gophers close by while the predator is still at a distance, again without alerting the predator.

It could also be a way of showing preference for your own family. Ground squirrels tend to live close to their own relatives. If one of them sees a predator and warns its own family by whispering a message in ultrasound, instead of yelling in the regular range, then its family might get to escape and the ground squirrel family next door will become lunch. Evolutionarily, that makes sense; it saves your own genes. And, certainly, the squirrels within range respond to this ultrasonic call. In the field, Dr. Wilson has seen one gopher silently call and the nearby animals scatter in response.

It also looks to be some kind of early-warning system. Typically, the ultrasonic calls get made only when a predator is off in the distance. Once it gets in close, then it's time to yell. No point in being subtle if you and your family are about to be eaten. Just scream your head off, and hope that everyone gets away.

Ultrasound is nothing new. Bats use it, certain whales can use it. But as an early-warning system, this is a novel use. At least it proves that the ground squirrels haven't just lost their voices from yelling.

CRAFTY CHAMELEON CAMOUFLAGE

If there's one animal that has a natural advantage when it comes to playing hide-and-seek, it's the chameleon. These low-profile lizards famously have the ability to change their skin to match their surroundings in order to avoid the eyes and mouths of hungry predators. It's a pretty impressive ability, not to mention challenging, as chameleons have to avoid a number of different predators, and not all of them see the world in the same way. Dr. Devi Stuart-Fox, an evolutionary biologist from the University of Melbourne in Australia, has been studying the Smith dwarf chameleon (*Bradypodion taeniabronchum*) and found that it adapts its way of hiding, depending on who's coming in to attack.

Smith dwarf chameleons, as their name suggests, are small, with a body length of about five centimetres (two inches). One can fit in the palm of your hand. Then there's the tail. It's about as long as the body and, like the tail of a monkey, can be used to hang on to tree branches. Unfortunately, this diminutive reptile is a critically endangered species, found only in one small mountain range in the southeast part of South Africa, so you're unlikely to see one in the wild. And, of course, they do have the ability to hide.

The changing of colour that a chameleon can accomplish relies on two factors. The first relates to the cells in the skin. These specialized cells each contain a different colour pigment, allowing the chameleons to turn their skin many different hues. The second factor is the role of the brain. These pigment cells in the skin are under the control of the nervous system. So, the animals can send messages to the pigment cells, causing some to open up, showing their colour, and others to close down and conceal their hue. *Voilà!* The classic chameleon quick-change.

How specific are these colour changes, though? Well, the chameleon has to deal with a variety of predators, so it may need to have a variety of tactics in order to hide. The eyes of a bird such as the shrike, which likes to feed on chameleons, are very different from the eyes of a snake. That's both a physical difference – birds have four cones in their eyes, while snakes and humans have only three, meaning birds can see a greater colour range than snakes – and an environmental difference – birds are looking down when they're hunting for chameleons, so the background is the ground, while snakes look up, and see chameleons against the bright sky. That specificity is what Dr. Stuart-Fox wanted to figure out.

She started out with a fairly straightforward experiment. She took a model of a shrike (the bird) and a model of a boomslang snake out to the native range of these reptiles, where she would show the chameleons one of the two models. She would either swoop down the bird, which she'd put on a stick, or brandish the model snake. "The first thing the chameleon does," Dr. Stuart-Fox says, "is flatten itself and flip to the opposite side of the twig. So it puts the twig, which it's on, between itself and the predator, and it changes colour. Usually they match their backgrounds quite well, but it changed colour subtly and in different ways in response to these two predators."

This is the interesting part of the experiment. When the predator was a bird, the match to the background was very good. But when it was a snake, then the chameleon would just become paler. This makes sense if you think about it from the chameleon's perspective. The bird, as mentioned earlier, has excellent colour vision, so, the chameleon changes its colours to match the background. But the snake isn't as good at identifying colours, plus it's going to be looking at the chameleon against the sky, so the chameleon makes itself invisible against the sky by lightening its body colour when it senses a snake approaching.

This response seems to be completely automatic for the reptiles. It doesn't matter what angle a snake or a bird attacks from, the response of the chameleon is always the same: shifting to the opposite side of the twig, then producing a good colour match to the background for birds and lightening itself for snakes. Which, for Dr. Stuart-Fox, was quite a surprise. "The differences," she says, "are quite subtle. And it's a remarkable ability to suddenly change your camouflage in response to two different predators."

We wonder what would happen if both the predators attacked at the same time.

5

PARSING PARASITE PECULIARITIES

Yes, they're gross and slightly unnerving, but even parasites have some intriguing behaviours...

BLISTER BEETLE DECEPTION

Nature loves throwing down challenges, and some of them don't seem very fair. Pity what faces the blister beetle (*Meloe franciscanus*) larva, a denizen of the Mojave Desert in the southwestern United States. Each beetle larva is about two millimetres long (less than a tenth of an inch) and a dark reddish-brown colour. It emerges from an egg that its mother laid under a desert plant and climbs up a stem. But the beetle larva doesn't eat plants. It needs to find its way to the nest of a solitary bee (*Habropoda pallida*), where it will feed on the nectar and pollen the bee collects while the larva develops. That could be on the other side of the desert. So, how is the larva going to get there? The sand is going to be about 50 degrees Celsius, and, as Leslie Saul-Gershenz, the

Director of Conservation at the Center for Ecosystem Survival in San Francisco, puts it, "They would literally fry if they tried to walk across the sand."

Finding one of these nests isn't going to be easy. As their name suggests, solitary bees live alone, building their nests in the sand. If you want to find your way there, what's the best way? Getting a bee to take you would make sense.

And so, that's what these larvae do. "What they do," says Ms. Saul-Gershenz, "is they literally call an air taxi – but it's not a direct flight." Here's where it gets cool. After the larvae crawl up onto the plant, they gather together. As many as a thousand of these larvae all clump together, in one big aggregate. While they might not look like a bee at this point, the lump they create is about the same size as a bee. Which is important, because the next thing they do is emit a chemical that mimics the sex pheromone of the solitary bee. Many insects react more strongly to smells than they do to visual cues, and the bee is one of those. A male bee flying by will smell the pheromone and think, "Aha! There's a female bee here, and she's ready to mate." So down he flies, and, well, it's time to copulate.

Imagine the scene. A male bee, ready to mate, flies down, lands on what he thinks is a female, and begins to engage in sex, or rather, as Ms. Saul-Gershenz calls it, performs "pseudo-copulation," since there isn't really any female there to have sex with. The larvae don't mind, though. They stop pretending to be a lady bee and climb onto the male bee's back. Here's where another evolutionary adaptation steps in. These larvae have incredibly well developed legs that allow them to grab on to the bee and stay attached.

Ms. Saul-Gershenz describes what happens next to the male: "I think he's probably a little surprised because it's not quite the experience he had looked for. Often he falls to the ground, onto the sand, rolls around a bit trying to get some of them off, probably because he can feel the weight of all these larvae on him. But he manages to get up again and flies off to look for another real female bee."

Now the larvae are on their "air taxi" – but their journey is not over yet. Male bees don't go to the nests. For that part of the trip, the larvae need to hitch a ride on a female bee. Luckily for them, the male bees might be burdened with the larvae, but they're still interested in mating. So, off they go, looking for a female, and when they find one, the larvae make the transfer while the mating is going on. Now they're set. Off goes the female to her nest to lay an egg (these bees lay only one at a time), and along with her go the larvae.

Journey complete, the larvae deplane and settle into the nest, where they spend the next several months feeding on the bee's larder. It's possible they're eating the bee's own offspring, too, although that's yet to be proved. Once they've gone through all their development stages, these beetles emerge from the nest, ready to start the cycle all over again.

There's not much in it for the bees, but you have to marvel at a system like this. Deception, hitchhiking, and food theft. Sounds more like a movie plot than biology, but that just shows how complex the natural world can be.

Ants Look Berry Nice

For most animals, it's a good idea to not look like food. They are probably better off, and safer, disguising themselves as rocks, grass, sticks, leaves, or not at all. This is why a group of researchers studying ants (*Cephalotes atratus*) in Panama were puzzled by what they saw. Some of the ants they were studying looked very appetizing indeed. They looked like a juicy and colourful treat for birds, and not only that, they were advertising the fact. Clearly, there was a mystery to be solved, and Dr. Steve Yanoviak, a tropical insect ecologist at the University of Arkansas in Little Rock, was part of the team that figured out just what was going on.

It wasn't their colour that initially attracted Dr. Yanoviak and his colleagues to these ants. They were originally looking at the ants' gliding behaviour. These particular ants live in trees, and if they fall off, they can orient themselves in a way that allows them to glide back to the tree's trunk and get back to safety. It was while studying this behaviour that one of Dr. Yanoviak's collaborators, Dr. Robert Dudley, noticed that some of the ants were red, but only some. At first they thought that maybe they'd found a brand new species. But then Dr. Yanoviak took some red ones back to his lab and, he says, "opened them up under the scope, and out poured a bunch of nematode eggs."

Nematode is the proper name for a roundworm. And while some of them live in the soil, there are plenty of nematode species that live as parasites on other creatures. Dr. Yanoviak figured that was what he was looking at here. The effect of the nematodes on the ant was quite dramatic. The eggs are found in the gaster, or hind section of an ant, and somehow they either make the exoskeleton thinner or they remove the pigments from the exoskeleton. The result is that the ant gets a shockingly red butt. "Especially when an ant walks into the sunlight," says. Dr. Yanoviak, "it really shows up well. Also, given that these ants tend to walk around with the gaster in an elevated position above the body, then it's very conspicuous when you see them."

So these ants are walking around with their butts stuck up into the air. As odd as that may sound, it's a typical posture, called gaster flagging, for ants when they're upset. And it seems these ants are really disturbed. In uninfected ants, about 1 per cent will be gaster flagging at any time. In the infected ones, the opposite is true, and 99 per cent stick their hind ends up.

Which makes them a very conspicuous prey item for birds. And this is the final clue in the story of the ant and the parasite.

Their red butts make the ant look like a berry, which is going to make them attractive to passing birds. The birds then bite off the ant's gaster (an activity confirmed by Dr. Yanoviak when he tethered ants in the field to see what would happen when a bird flew by) and eat what they think are berries. The nematode eggs, contained in the ant, pass through the bird and end up in its feces.

Now, as it turns out, these particular ants happily take bird feces back to their nests as food for their larvae. (As a side note, bird feces contain lots of protein from undigested insect parts, and plenty of salts and minerals too, so they make a great food for growing ant larvae.) Unfortunately, if the feces also contain the nematodes, these will end up inside the larvae, where they migrate to the developing gaster. Dr. Yanoviak picks up the story: "Just around the time that the ant emerges as an adult, those adult nematodes in its gaster are mating. And the ant is entirely black at that time. Now, new adult ants don't forage outside the colony very much. They mostly stay inside tending to the brood. As the ant gets older, it starts to go outside more. At the same time, the nematode eggs that were caused by that mating of the adult nematodes and the ant are now starting to mature. And the ant is getting redder as time goes by."

It's a pretty good system for the parasite. They never have to worry about wandering around, hunting for food. Just by manipulating their host, they get a free ride and a chance to spread their genes around. Not so much in it for the ant, though . . . but then, who said evolution was fair?

CRUSTACEAN KIDS, PARASITIC PARENTS

Imagine a planet populated only by babies. Everywhere you look, you see infants but no adults. This would be very mysterious. Where did the babies come from? What will happen to them when they grow up? Of course, this has never happened to humans, except perhaps in an old episode of *The Twilight Zone*, but it's been a mystery surrounding one group of crustaceans for a very long time. These animals have only ever been discovered in their larval form, called a y-larva or facetotectan. No biologist has ever identified the adults of the species. That is, not until Professor Jens Høeg, a biologist from the University of Copenhagen. Using clever lab techniques, he thinks he's cracked the conundrum. However, he's opened up as many questions as he's answered.

Y-larvae have been known about for more than a hundred years now, and as crustacean larvae go, they don't look unusual. They're typically small, like other crustacean larvae, measuring less than half a millimetre (two hundredths of an inch) long. They're also in all the oceans, from the sub-Arctic waters off Scandinavia and Greenland to the near-tropical waters of Okinawa in Japan. The team studying them in the coral reefs off Okinawa have found more than forty different species. That's forty different species, and not a single adult form found for any of them.

The form they're easily found in is called the nauplius stage. The name comes from Greek and means "one-eyed," and, yes, they have only one eye. The nauplius stage is something every crustacean goes through – and, except for this group, they keep developing into their adult form. But not so with the y-larvae. You can take these into the lab and they just remain larvae, as if to confuse the researchers.

One clue appeared in the 1960s, when a Danish scientist was able to coax some of the nauplii to develop into something that looked a little like a shrimp. They weren't adults, they were simply a bit more advanced than the nauplii, but it was a start. What the researchers found curious about this shrimp-like organism is that it seemed to be specifically adapted to attach onto something, as if this organism needed to latch itself onto the coral, or the seabed, or perhaps another animal.

Fast-forward to Dr. Høeg's lab experiments. "First we tried various things," he says. "We tried to rear the larvae and see if we could get them to develop into something just by ordinary culture techniques. That didn't work. We searched for the adults in the local environment. That didn't work, either. And then we thought of the trick of trying to induce them in the laboratory to change into the later stage by giving them various chemicals."

A bit of background here. Crustaceans, like their relatives, the insects, go through various stages, or moults, on the way from juvenile to adult form. Each of these involves the production of hormones that signal the moult to start. Then, the crustacean loses its old exoskeleton and a new, larger one forms in its place. It was these hormones that Dr. Høeg was using. And, sure enough, he found one that worked. It was a hormone produced by crabs and shrimp to trigger the moult from juvenile to adult form, and it seemed to do the same to the y-larvae.

"That was the exciting moment," he says. "We had these larvae that looked something like shrimp, like most people would say. We gave them the hormone, and within a few hours, a little sloth-like or worm-like thing that looked nothing like a crustacean crawled out of this larval skin. It just looked like an unstructured worm, with just a few types of cells inside. No appendix, no legs, no sensory organs. No mouth. Nothing."

This was astonishing, and certainly not what most biologists would have predicted for these animals. But it does give a partial answer. These "unstructured worms," as Dr. Høeg calls them, have been seen elsewhere in nature. They're a typical form of parasites. So these free-swimming larvae are actually the offspring of some parasite that's living inside some other creature.

What that creature is, or, more likely, what those creatures are, is still a complete mystery. No one was expecting to discover these larvae were from a parasite, so no one's looking for the adult version yet. Even what triggers the transformation in nature is a mystery. It's probably the case that the larvae encounter their eventual host, which triggers a moult, but no one knows for sure. There's lots more work to be done.

The fact that these larvae are so common in the ocean suggests that they're playing an important role in the ecosystem. And while we used to think about all the bad things parasites do to their hosts, more recent biological thinking is that there are both benefits and burdens to carrying another organism around with you.

So, there's the next mystery. Scientists know what they're looking for, now they just need to find it.

Woolly Caterpillar Medicine

Being a caterpillar can't be easy. Sure, there's the chance of turning into a beautiful moth or butterfly, but to get this reward, there's at least one grim risk. On the road to the cocoon, caterpillars have to deal with parasites – flies and wasps whose offspring set up shop inside them and kill the host in the most unpleasant manner (think of the movie *Alien*). But, according to Dr. Michael Singer, a biologist from Wesleyan University in Middletown, Connecticut, the caterpillar of at least one species has come up with a way to fight back. The woolly bear caterpillar (*Pyrrharctia isabella*) practises a form of traditional caterpillar medicine to treat the invasion.

The woolly bear caterpillar is found throughout North America and is the larval form of the Isabella tiger moth. They're fairly easy caterpillars to recognize: the head and tail are black, but in the middle of the body, the hairs are an orange-russet colour. Woolly bear caterpillars hatch in the fall and survive through the winter before emerging as butterflies in the spring, and folklore says that the more black there is on the caterpillar, the worse the upcoming winter is going to be (though science doesn't back this up). Unlike many other caterpillars, the woolly bear has a really varied diet, which comes in handy when dealing with parasites.

Parasitism is problematic for these creatures. Flies and wasps, which are correctly referred to as parasitoids, lay their eggs either on or inside the caterpillar. Once the eggs have hatched, they set up what Dr. Singer refers to as "a feeding site," and feed on the blood of the caterpillar. Then comes the emergence of the parasitoid. This is the *Alien* part, so skip to the next paragraph if you're squeamish. The parasitoid has now finished its development and needs to get out of the caterpillar's abdomen. It takes the most direct route possible, through the side of the body. No, the caterpillar doesn't survive this. As Dr. Singer says, "It seems like a miserable way to go."

But the woolly bear caterpillar doesn't just roll over and accept its fate. It tries to kill the parasitoids before they kill it. Dr. Singer's first clue to this was the realization that they were eating plants whose leaves contain toxins. This is odd, to say the least. Humans might eat things that are bad for them, but most other critters are smarter than us and avoid toxic foods. Yet the caterpillars were munching down on leaves that might kill them.

Dr. Singer then discovered that it wasn't until a caterpillar has been parasitized that it starts eating the toxic plants.

It's not as crazy as it sounds. It turns out that the caterpillar is able to produce an enzyme in its bloodstream that neutralizes the toxins. These particular toxins produced by plants are called pyrrolizidine alkaloids, or PAs, for short. PAs are produced by plants as a defence against insects, but this insect has adopted the system for its own use. It neutralizes the PAs, which build up to a really high concentration in the caterpillar's bloodstream. Then, the parasitoid drinks the blood of the caterpillar, but it doesn't have the enzyme. The PAs turn back into a toxic form, and the parasitoid is done for.

Of course, this is a careful balancing game. If the caterpillar ingests too much toxin, it's not able to neutralize it all, and it ends up as one dead caterpillar. This does happen. Sometimes, in an attempt to protect themselves, the caterpillars overdose. But the risk is worth it. When Dr. Singer looked at the effect of self-medication on survival, it improved the caterpillar's chances by 20 per cent – which, at the population level, is a very significant effect.

This isn't the only example of self-medication in the animal world. Plenty of vertebrates do it too. But no one ever expected to see this kind of advanced behaviour in an insect. Time to start looking a little more closely at these hard-shelled invertebrates.

6

HIGHER, FASTER, STRONGER

Humans are rightly proud of their athletic accomplishments, but few can hold a candle to most animals. They run faster, throw farther, and see better than we do, although some of their records are ones we might not want to try to beat . . .

THE FASTEST JAW IN THE SOUTH

The world needs a new superhero. Let's call him "Ant Man." He leaps high into the air, has a grip like a steel trap, and boasts the fastest jaw this side of the Rio Grande. Sounds preposterous? You don't think those are the right superpowers for an ant? Well, think again. The trap-jaw ant (*Odontomachus bauri*) is able to leap several times its own height, catapult itself backwards through the air, and has the fastest snap of any species ever found. Dr. Sheila Patek, a biologist from the University of California, Berkeley, is part of the team that timed the ant, and has given it the label "Top Jaw."

The trap-jaw ant is found throughout Central and South America. It's a comparatively large ant, with a body length of about one centimetre (slightly longer than one third of an inch). Its most recognizable feature is the jaw. It has two mandibles (the proper name for jawbones), which stick out horizontally: imagine a pair of C-clamps oriented parallel to the ground, sticking out from the front of the ant's head. The ant closes its two mandibles in front of its face and opens them wide, almost at ninety degrees to its body, like a bear trap, ready to swing shut.

It's this swinging shut that's interesting. Dr. Patek first became involved in studying these creatures when a colleague couriered her a package of ants and asked her to measure how fast they could close their mandibles. Dr. Patek is one of the world's leading experts on speedy critters, so she was the obvious choice. She took the ants, mounted one under a microscope, and, she says, "hooked up an extremely high-speed video camera to the top of the microscope. This video camera can film at up to 250,000 frames per second, which is really, really fast. So I hooked up the ant, put her under the microscope, and I bugged her a little bit so she would snap her jaws."

And boy, did those jaws move fast. The jaws were closing at between 35 and 60 miles per second, which translates to an average of about 126,000 kilometres per hour, or an acceleration of over 100,000 times the force of gravity. "That's a huge acceleration," says Dr. Patek. "But I'm looking at these numbers and I'm thinking, if we banged our arm against the wall with [a force] three hundred times our body weight, we would either break our arm or push ourselves backward." That really got her thinking. Obviously these ants are able to accelerate their jaws to incredible speeds, but what forces are they generating, and then, what do they do with that force? Now the research moved beyond finding a speed record, which they had, to study how that speed was applied.

The first clue to what was going on emerged from papers written back in the 1800s. Back then, researchers had noticed trap-jaw ants jumping with their jaws. Then, in the 1980s, another paper came out by a biologist who had, according to Dr. Patek, "witnessed these ants firing their mandibles against intruders and actually throwing intruders away from them. So, it was like a huge punch from the mandibles that caused the intruding ants to fly backwards twenty centimetres [nearly eight inches] or more."

Dr. Patek got her collaborators, Dr. Andy Suarez and Dr. Brian Fisher, to send her more ants from Costa Rica, and she started examining what they were doing with their jaws. The experiment was quite amusing. Dr. Patek says, "We put some kind of intruder in the arena with the ant, like a big spider or another species of ant, and sometimes the [trap-jaw] ants decided to attack the intruder, but sometimes the ants think, *I just want to get out of here.* In that case, they angle their head towards the ground with their mandibles cocked and open, they wiggle their body around, sometimes they lift one leg up or the other, they get everything into position, then they fire the mandibles against the ground, which causes them to fly into the air, spinning dramatically."

What goes up must come down, so these ants do have to land. But because they're so light, they float back down and don't seem to suffer any harm. Once they land, they just run away, avoiding the original danger. Their jump is impressive, though. They're able to get about eight times their own body length off the ground. Not bad for a vertical leap from standing still! And all powered by the jaw. Good for comic relief, too. "Oftentimes in the lab," says Dr. Patek, "the ants would get riled up and they would get annoyed with us and all of a sudden, there would be ant popcorn everywhere, ants flying all over, landing on you, sitting on you. So it's bizarre."

And that's not the only way these ants use their jaws – they also do something Dr. Patek calls "the bouncer defence." When an intruder gets too close, and the ants have decided not to make a jump for it, they'll approach the other creature and fire their mandibles. The force pushes the ant itself back, but, just as Newton's third law of motion would predict, the other animal is pushed backwards as well. If the intruder is small, it takes the brunt of the force and is knocked back out of the way. If it's something big, like a spider, then it's the ant that flies back – but now it's well away from the potential predator, in this case up to twenty body lengths.

And you thought "Ant Man" was a silly model for a superhero!

Preying Mantis Shrimp

Back off, Bruce Lee! Watch out, Mike Tyson! You might think these fighters have fast hands and a powerful punch, but neither of them would last a minute in the ring with nature's record-holding mantis shrimp. The mantis shrimp are actually neither true shrimp nor mantids. Rather, they're a crustacean that's closely related to the shrimp. This small but fearsome predator can punch with its forelimbs faster and harder than any other animal, including humans. But, as Dr. Sheila Patek discovered (she's the researcher at the University of California, Berkeley, who told us about the amazing trap-jaw ant), it "cheats" to do this, with a mechanism not seen elsewhere in nature.

The mantis shrimp isn't much like the kind you might see on your dinner plate. First of all, it's much larger, with some species, including the one Dr. Patek studies, as long as fifteen centimetres (six inches). And it has very long legs, which look like the ones on a preying mantis. They give the animal its name. These legs are also the key tools for the mantis shrimps when they hunt. In some species, the shrimp hide in burrows and reach out to jab and stab their prey. This could be considered the Mike Tyson approach. Other mantis shrimp can jab, but have a second, more interesting approach. Dr. Patek calls it "smashing," where the shrimp slams down its arm, "sort of like smashing with your elbow." More like the Bruce Lee approach to breaking bricks with his forearm.

Smashing shrimps have been known about for years. It was by accident that Dr. Patek discovered their record-breaking behaviour. She'd originally been studying the sounds of these shrimp, which she compares to the sound of "popcorn popping," and decided to look more closely at their front leg movement. Here's how she describes what happened: "I decided to take some high-speed video, a special kind of video that can slow things down a lot. And when the three of us that are on this paper got together and started to film with the cameras that we had available, we actually could not see the strike. It was just too fast. Then, through a collaboration with the BBC, we were able to rent a new type of high-speed video that could film at very high resolution, at 5,000 frames per second, so we could actually slow down the strike 333 times and see what's happening over the course of that strike. And it was when we slowed it down that we realized that they were moving at extraordinary speeds."

Those extraordinary speeds were about ninety kilometres (fifty-six miles) per hour. That's pretty fast on its own, but then think about where the shrimp is moving the arm. This is all happening underwater, and as anyone who's tried running in water will tell you, there's a lot of resistance there. Somehow these shrimp are generating enough force to move their limb through the water at speeds fast enough to smash the hard shells of their favourite food: snails.

Not only is the movement faster than anyone expected, but it also uses a mechanism no one's seen before. One part of the secret to their speed was uncovered over forty years ago. The limbs have a latch. When the limb is pulled back (think about bending your elbow), the "wrist" catches on the upper arm. Then the shrimp contracts its muscles, storing up energy, and when the latch releases, the arm springs forward. But when Dr. Patek examined this in detail, she couldn't see how any of the muscles could have produced enough energy to cause the spring forward they saw. Not till they examined the latch mechanism more closely. That's where the saddle spring comes in.

This is one unusual structure. Imagine a saddle like one for riding a horse, only a lot smaller, of course. Now, visualize pulling the sides of the saddle down and in, then pulling the front and the back of the saddle up and together. That's what the shrimp does. It has one of these structures on each leg: the muscles compress the sides down and in, and the front and back up and in, creating a huge amount of force. When the latch releases, this saddle springs back to its original shape as well, and the leg shoots forward to smash into the prey.

All in all, its a remarkable device. But that's not the only amazing thing about the mantis shrimp. Remember the popcorn sound mentioned earlier? Well, that's actually a phenomenon called cavitation. As the limb thrusts forward through the water, it goes so fast it creates a small air bubble behind it. The friction in the water is so strong that it heats the water behind the leg to the point at which it evaporates. Then the bubble collapses just as the leg gets close to the prey. That generates heat and light, and, as Dr. Patek puts it, "It's basically a little explosion that's happening right next to the surface of the snail." Pop goes the snail from this double whammy of an explosion and a hammering.

The moral of this story? Handle these shrimp with care. They might be small, but a smash from these creatures will leave your hand throbbing.

The Mantis Shrimp's Super-Sight

For years, the majestic eagle has held a proud place in the animal kingdom for its sharp vision. That's why we refer to someone with keen sight as having "eagle eyes." But the eagle has some unlikely competition. The animals with really top-notch visual abilities, according to Dr. Sonja Kleinlogel, a neuroscientist with the Max Planck Institute for Biophysics in Frankfurt, Germany, are the mantis shrimp. Besides the remarkable punch they can pack, as we've just seen, these coral-dwelling crustaceans are at the top of the evolutionary heap for vision.

Their eyes are big. "The head is tiny and the eyes are huge," says Dr. Kleinlogel. "One eye is bigger than their whole head. It's basically two eyes hanging on a tiny brain." But the fun doesn't stop there. "Both eyes move independently," she says. "So when you meet a mantis shrimp, you always feel like they're watching you. They look sort of human-like, and not like a shrimp that just does its own thing. They're really weird to look at."

Oversized eyes that move independently – that sounds like the perfect system for a neurobiologist to study, and that's what Dr. Kleinlogel and her colleagues have done. What they've discovered is complex. We'll take it one fact at a time.

First of all, these are compound eyes, like an insect's eye. In some of these species, there are more than ten thousand of these individual eyes (called ommatidia) working together. But unlike an insect's eye, the eye of the mantis shrimp is divided into three regions. There are two flattened hemispheres separated by a third region, a band of ommatidia that runs horizontally around the eye. Dr. Kleinlogel says, "Each of these areas has its own pupil. So there are three black pupils looking at you." Think about the consequences of that. Each eye has multiple pupils that are separated in space from one another, which is pretty much what we have with our two eyes. In other words, each eye of the mantis shrimp is capable of stereoscopic vision, which means one eye alone has depth perception. It's almost impossible to imagine what that would be like when you take into consideration that there are two eyes, both with stereoscopic sight, on stalks, which could be looking in completely different directions.

If that's not astonishing enough, there's the matter of what they're able to see. Dr. Kleinlogel and her team started looking at the photoreceptors in the eye. In the human eye, we have photoreceptors for three different colours – red, green, and blue. (We also have rods, which are photoreceptors for detecting levels of light.) The mantis shrimp? "These guys have up to sixteen photoreceptors," says Dr. Kleinlogel. "So they can see different wavelengths, from the ultraviolet up to the far red." This obviously gives them a much wider range of colour than anything we can see, so it becomes hard, but not impossible, to imagine the world seen with a mantis shrimp's eye.

But there's still more to the story, because there's one type of vision that mantis shrimps have that we can't even begin to imagine. These animals can sense the polarization of light. We can get close to understanding this if we think about what happens when we wear polarized sunglasses. Natural light comes towards us from all angles – horizontal, vertical, and everything in between.

When we put on polarized sunglasses, we cut down the range of light so that we see only those light rays coming at us from the horizontal. That means the colours become sharper, the glare is reduced, and we can even see through the surface of water that's reflecting sunlight. Mantis shrimps can detect polarization, but not just the one way we do with sunglasses. They're detecting four different angles of polarization, and, says Dr. Kleinlogel, "On top of that, they have two which are sensitive to circular polarized light. This is light with a wave circling around in a circle. And that's something we can't even imagine yet. Machine vision can do it, but we can't really imagine what it looks like." So we're into a world of vision beyond anything we can experience.

Why do these creatures have such a complex visual system? The jury is still out, but Dr. Kleinlogel thinks it has to do with living underwater. Underwater, the red wavelengths start to disappear, and, to us, everything starts to look blue. These animals, by having a wide range of colour receptors, can adjust their vision to cope with the changing range of colours available as they get deeper underwater. Also, by having such strong polarization correction, they are able to keep images sharper as they drop lower down.

This could provide a number of advantages. Certain prey species of the mantis shrimp reflect polarized light, so having receptors for polarization could help in hunting. The enhanced colour vision might be useful for signalling potential mates or enemies, and to avoid predators, while improving the ability to hunt. For a tiny-brained animal like the mantis shrimp, the more information it can process in the eyes, the better it's going to do. Its complex eyes, able to process a lot of information very quickly, likely make it easier for the mantis shrimp to catch swiftly moving food, like fish.

However it came about, it shows that there are plenty of different ways to see the world around us.

Mandible Mayhem

The mantis shrimp isn't the only creature with a terrific punch. Dr. Jeremy Niven, a neuroscientist at Cambridge University in England, has discovered that the soldiers of the Panamanian termite (*Termes panamensis*) can hit faster than any other animal out there. A single blow from a termite mandible can kill its enemy.

How fast? Well, the record-breaking strike is 70 metres per second, or about 250 kilometres (155 miles) per hour. Imagine trying to move your arms at that speed, and you get an idea how fast these insects are operating. Of course, a termite is a lot smaller than a human, so the distances their mandibles have to travel are very short, and the whole action lasts about one-forty-thousandth of a second. Very fast, and very short – that's how to characterize the mandible motion.

The mandibles are two spikes that stick forward from the front of the termite's head, and are actually its jawbones. The spikes curve slightly inward so that they touch each other at the tip, in front of the termite's mouth.

These soldier termites exist solely to protect the nest. They'll position themselves in the many tunnels leading down to the nest's heart, ready to take out any invaders. To do this, they need to prepare their mandibles. First they pull them back, then they pull the tips apart, and, much like a loaded crossbow, the mandibles are full of energy, ready to spring back. Then, when the termite encounters an enemy, it lets the mandibles go and they snap together. Since they're flying at each other so quickly, if the two mandibles met, they would be smashed. So they don't. Instead, they slightly overlap at the tip, and they snap together in the same way scissors do when they close.

If the termite has good aim, the mandibles are directed to a point just behind the head of the unfortunate intruder, roughly where the neck would be if the enemies were mammalian. But we're talking about insects here, and the spot where our throat is situated is where the nervous system of an insect comes together. Hit this spot and it'll have a dramatic effect. The punched invertebrate will stagger around for a few moments, but then it will collapse, dead. This is a blunt-force trauma; the termite never pierces its adversary's outer surface. The internal damage is severe, though. It's also quite effective in defending the colony. Only about 1 per cent of the termites will be soldiers, presumably because they don't need any more when they have such a weapon up their sleeves.

Want to see this for yourself? You're going to need a very good camera. When Dr. Niven started researching this, he had a high-speed camera that could shoot a hundred frames a second. "All we saw was a blur," he says. So, they started shooting at faster and faster speeds, and, he says, "We had to go up to forty thousand frames a second before we finally saw it." This may be the fastest-moving limb found anywhere in nature.

Luckily for us, we're a lot larger than the typical enemy of the termite. But if you're digging for them, "You can feel them," says Dr. Niven. "If you hold them on your hand they'll occasionally do one of these mandible snaps, and you can feel it." So, if you're in Panama and you feel the bite of a termite, show respect for its power and speed.

How the Bow Bug Shoots Itself

Imagine, if you will, an insect whose survival strategy is to shoot itself. It doesn't use any tools, but these insects shoot themselves in the same way a bow shoots an arrow – only they're not just the arrow, they're the bow as well. If this sounds confusing, well, it is. But Professor Malcolm Burrows and Dr. Greg Sutton, both from the University of Cambridge, England, have figured out how it works. By looking at the biology and the physics of the froghopper (*Philaenus spumarius*, also known as a spittle bug), they've worked out how these insects project themselves up, up, and away.

Froghoppers are reasonably common bugs found all over the world. They survive by sucking juice out of the leaves of plants, and the young bury themselves in a sticky mass stuck on the plant that looks like spit, hence the name spittle bugs. But froghopper isn't such a bad name either, as Professor Burrows admits. They do look a little like a frog (although much smaller at about six millimetres, or a quarter of an inch, long), and they certainly can hop, up to about sixty centimetres (two feet), which is a very impressive jump given how heavy they are. That's jumping one hundred times their own body length, or, for humans, jumping straight up more than one hundred metres or yards (two hundred if you're a basketball player).

These jumps aren't just high; they're fast, too. If you don't have a high-speed camera, you're never going to catch one in action. The whole process lasts one fiftieth of a second. Dr. Sutton explains, "If you blink, you will miss fifty jumps." That's a lot faster than the eye can see. The researchers only discovered this incredible speed by accident. "We first came across this on a field course," says Professor Burrows. "We had just a rather poor camera with us, and we put the bug in front of it, switched the camera on, and then recorded what we saw. And we had one frame with the bug sitting there, and then the next frame there was nothing."

This prompted them to get a much faster camera that could take eight thousand frames a second, and work out the takeoff speed. Once they had that, they could calculate the G-force of takeoff. The results? Brace yourself: these bugs are experiencing

400 G. That's four hundred times the force of gravity. We pass out if we go beyond about 5 G, and on the space shuttle it's limited to 3 G. We've got nothing on these bugs. They're going fast, they're going high, and they're doing it under a lot of pressure.

Not only could the fast camera tell the researchers the speed of the bug's movement, it also allowed them to see which joints were involved, and how fast the limbs were moving. They knew right away that the muscles couldn't possibly contract quickly enough to power this jump, so they had to look for another mechanism. That's where the bow and arrow idea comes in.

We'll start by describing the bow. You can't see it by looking at the insect; it's on the inside, in the thorax, which is roughly where our chest is. A quick bit of biochemistry will help here. Insects produce an elastic protein called resolin. It's easy to see, because under ultraviolet light it's fluorescent. When the researchers put a froghopper under the microscope, Professor Burrows says, "Here were these two great headlights, which shone out from its thorax, of bright blue. The longer we looked at it, the more elaborate and more extensive this resolin became. It was in a bow shape. And when the legs are primed to make the animal jump, you can actually see this bow bending."

That was a pretty remarkable discovery. Inside this insect's chest, there's a protein that's shaped like a bow, which, when the rear pair of legs starts to bend up in preparation for a jump, bends too, the same way that pulling back on a crossbow bends the prod or bow. But that's not all. When they looked at the physics, they found that this protein alone doesn't have the kind of strength and stretch the insect would need for takeoff. For that, there's got to be more there. A closer look revealed a layer of hard insect cuticle (the outer shell of the insect), which is bound together with the resolin as a composite material. It's only when both are there that there's enough strength and spring for the insect to push off at speed. Again, this matches what happens in a crossbow, which is made from composites of elastic and firm materials.

The resemblance to the crossbow continues. If you've ever pulled back on a crossbow, you'll know two things. One, it's hard to pull back. Two, once you've pulled it back, it's hard to keep it there. The froghopper has answers to both problems. To deal with the pull-back, it has a pair of large muscles inside the thorax that are responsible for tensing the bow. Like humans pulling on a crossbow, this is a reasonably slow process – it takes the insect a couple of seconds to complete – so it's not a move that the froghopper rushes into. But, once the bow is taut, the insect doesn't have to let go immediately. As with crossbows, there's a latch to hold the bow in place. Then, the insect doesn't have to release this pent-up force until it needs to escape in a hurry. And remember, these guys have six legs, so they can keep their last pair cocked for a long time in case they need to get away.

This system makes the froghopper the world record-holder when it comes to jumping. Other species might be able to jump as far, relative to body length, but if you look at body mass, nothing comes close. We can generate about twice our body mass in force. The flea can get up to about a hundred times. These guys are generating roughly four hundred times their body mass.

However, as good as this approach to transport is, you might want to think twice before you load yourself into a crossbow. You wouldn't do well at those kinds of accelerations.

Catapulting Caterpillar Feces

There are certain activities we all do every day. One is eating, but another, equally important activity is defecating. We've designed a sophisticated way of flushing away the waste, whisking it into sewer systems that take it out of sight and out of mind. But other species don't have this luxury. So one, the caterpillar of the silver-spotted skipper (*Epargyreus clarus*), has come up with an ingenious method of waste disposal. It fires its feces a metre or more away from its home. Dr. Martha Weiss, an environmental biologist at Georgetown University in Washington, D.C., thinks she knows why these critters go to so much trouble to propel their poop: it's all about preventing predation.

These caterpillars have quite the range for their propulsion. Dr. Weiss's lab has recorded a caterpillar shooting forty times its own body length, or, as Dr. Weiss says, "We have taken the trouble to calculate the fecal field-goal analogy, and if we had a six-foot-tall kicker, this would be the equivalent of his making a field goal from his own team's thirty-four-yard line, so that would be a seventy-six-yard field goal." That's better than anything you're going to find in the NFL.

If you want to avoid getting pelted by caterpillar poop while strolling in the forest, you'll be pleased to hear that these insects are easy to identify. They're found throughout southern Canada and the United States, and they're quite striking, so to speak. They're green, with yellow stripes, a brown head, and a red neck. While they're only the size of a grain of rice when they hatch, by the time they're ready to pupate they can be four centimetres (an inch and a half) long. Of course, even if you did get hit, it wouldn't be that bad. Unlike other forms of feces, caterpillar poop, or frass, as it's technically known, is simply leaf material that's all balled up. Not the kind of stuff that would leave much of a mark if you were pelted.

How these caterpillars fling their frass is quite ingenious. Their last body segment has a flattened, fan-shaped plate right above the anus, called an anal comb. This plate latches below the anus on to the final segment, which can then fill up with pressurized blood. So, the caterpillar gets ready to poop, then the frass comes out and sits on the plate. Underneath, the segment starts to fills with blood until it's almost bursting. Suddenly, the latch lets go, and the plate springs forward, catapulting the frass into the air at up to 1.5 metres (59 inches) per second. No fuss, no muss, and the feces have left the fan!

There are three possible reasons why this caterpillar goes to the trouble of firing off its frass. It could be the "hygiene hypothesis," that is, the caterpillar just wants to keep its nest clean, the same way that we like to keep our houses free of poop, and the bacteria and viruses that might come with it. Then there's the "crowding hypothesis," which speculates that it's not about disease but simply about space. If you leave all your frass lying around (and remember, these caterpillars live with all their siblings), then the nest would soon be full, which would mean building a new nest – an activity that's both energetically expensive and dangerous.

The third idea is the one Dr. Weiss thinks is the real answer. That's the "natural enemies hypothesis." The hypothesis is that frass can attract predators, either by its odour or appearance, and these predators may find a caterpillar and, as Dr. Weiss describes it, "either kill it to eat it or lay eggs inside of it so that their larvae could develop inside the body of the caterpillar." A fate any animal would try to avoid.

Dr. Weiss tested this idea by collecting colonies of paper wasps, a natural predator for the silver-spotted skipper caterpillar, and introducing the wasps to caterpillar nests, both with and without caterpillar frass inside. Sure enough, the wasps hung out wherever they found the frass. In one experiment, when the caterpillars were there too, then the wasps killed them. It seems that cleaning out the home improves the chances of avoiding attack for these caterpillars.

Mom was right after all: don't forget to flush. You never know who might be attracted by the smell of your frass – at least, if you're a caterpillar.

7

EVEN MORE WEIRD BEHAVIOUR

Some animal antics refuse to be categorized, but they do show us that there's plenty to learn by looking at creature behaviour, even if, occasionally, it seems a little disgusting.

Washing with Urine

Many aspects of animal behaviour attract the attention of scientists. The elaborate courting dances of birds, for instance, provide insight into reproduction. Studying the intricate schooling patterns of fish shows how simple behaviour can lead to complex, organized actions. And research on social dominance in baboons has given us insight into animal psychology. There are some behaviours, however, that have escaped explanation, perhaps because they are so bizarre that few scientists want to take on the challenge of teasing out their secrets. Take, for example, the question of why monkeys pee on their hands and feet, a practice delicately called urine washing by biologists. Primatologist

Dr. Kimran Miller (now known as Dr. Kimran Buckholz, and teaching biology at Wartburg College in Iowa) thinks she's solved this primate conundrum.

If you've never seen it, urine washing is pretty much what it sounds like, and many monkey species do it several times a day. Dr. Miller was studying capuchin monkeys (*Cebus apella*) when she decided to delve into this odd behaviour. What she'd noticed happening repeatedly was a monkey urinating on one hand and then wiping that hand on the foot on the same side of its body. Sometimes the monkey would then repeat the action on the other side of its body.

Something as weird as this hadn't gone unnoticed before Dr. Miller's work, but for all the different hypotheses out there as to why monkeys might do this, no one had ever put any to the test, which is what Dr. Miller did.

The first idea she tested was that this behaviour had something to do with temperature regulation. Unlike humans, monkeys don't sweat. When we sweat, the evaporation of moisture from our skin cools us down. The hypothesis held that monkeys were replacing the fluid from sweat with the fluid from urine, so that when it evaporated, they'd cool off. To see whether this held water, so to speak, Dr. Miller looked to see if urine washing increased on hot days. The result? Dr. Miller says, "We found no relationship between urine washing and climatic conditions." There goes that idea; monkeys aren't peeing on their hands to cool down.

The next leading idea was that it might have something to do with marking territory. This one's pretty straightforward. Lots of animals mark their territory with urine, as any dog owner will tell you. Maybe by peeing on their hands and feet, the capuchins were efficiently spreading their own scent around. Dr. Miller tested this by noting how frequently they urine washed when other groups of monkeys were around. If they were marking territory, they'd urine wash more when others were close by, to make sure those potential rivals got a clear picture of whose land they were trespassing on. Again, the results were clear. "We found that they actually urine washed more outdoors where there were no neighbouring groups," says Dr. Miller, "and so we ruled out the idea that they were urine washing to communicate territorial boundaries to other groups of capuchins." Strike two.

Other ideas came and went – using urine for improving the grip, for instance (like spitting on your hands before throwing a ball). But then she found the key. "We began noticing," says Dr. Miller, "that sometimes, when an animal looked at another animal, particularly the alpha males, that they would often urine wash afterwards." Not just that, but there seemed to be a link between that look and the subordinate animal becoming agitated or responding as if it were being threatened. Hundreds of observations later, the data supported the observations. "If an animal is displaced from a certain location in their enclosure," says Dr. Miller, "or possibly if they catch a glance of another animal, just any kind of subtle, mild aggression, they will often urine wash afterwards." Often, in this case, was about 87 per cent of the time. And it had an effect. When they urine washed, the aggression would stop.

Why this works to stop aggression is a mystery. One idea, says Dr. Miller, is that the monkeys are saying, "I get it. I'm sorry. I messed up. Here, watch me urine wash and this will make it all better." The ultimate way of showing your superiors that they're better than you.

Alternatively, this could be a calming mechanism for the monkeys. One of the observations she made is that cortisol levels in capuchins drop when they urine wash. Cortisol is a hormone that rises when stress levels go up. Dr. Miller thinks, "The behaviour could serve as a self-soothing behaviour that these animals perform to make themselves feel better."

The next time you're near the monkey cages at the zoo, you might want to think twice before shaking hands with the capuchin, especially if it looks stressed.

Spider Venom Side Effect

Getting bitten by a spider is rarely a good thing, and it will never lead you to be able to swing between skyscrapers, like Spider-Man. But in the case of the Brazilian wandering spider (*Phoneutria nigriventer*), there is a potent side effect. The bites from this little black critter are rarely fatal, but they do cause pain, discomfort, and a rise in blood pressure. Oh, and one little uplifting effect – literally uplifting, since these spider bites mimic that famous little blue pill. That's right, the spider bite comes with its own natural Viagra, and the effects can last for hours. Dr. Romulo Leite (at the time a physiologist from the Medical College of Georgia, now back in his native Brazil at the Federal University of Ouro Preto) has figured out how the venom makes men rise to the occasion.

These spiders are large creatures, they're native to the tropics, particularly Brazil, and, when spread out, they can measure up to fifteen centimetres (six inches) across, the size of your hand. They come in a variety of brown shades, from beige to black. Luckily for us, they only rarely show up here in Canada or in the United States. When they do, it's because they've hitched a ride on a bunch of bananas, which is where they like to live. In fact, some people call them banana spiders.

They're quite aggressive beasts. Unlike most spiders, which hang out on their webs and wait for their prey to come to them, these arachnids are hunters. They'll approach their prey and bite it to inject their venom. Which means that people working with banana trees will, from time to time, get bitten by one of these spiders. And that's when the problems begin.

From a doctor's perspective, the tumescence that results from the bite isn't the first concern. Minimizing the pain and making sure the venom doesn't kill the victim (it causes muscle paralysis and breathing difficulties, which can be fatal in children, although less dangerous in adults) are the initial issues. Then they have to make sure the blood pressure doesn't spiral out of control. It's only researchers like Dr. Leite, a specialist in erectile dysfunction, who want to figure out just why and how this bite causes uncontrolled erections in the victims. Erections, by the way, that victims complain about, since they can last for more than an hour, and can be painful.

In order to study this, Dr. Leite first separated the different chemical components of the venom, then injected each one under the skin of mice and rats to see what happened. Sure enough, he was able to identify the culprit chemical by its expected effect on the animal. Remarkably, he could induce erections in mice even when they were fully asleep. It's powerful stuff.

What he's found is a small protein molecule, called a peptide, which causes the release of nitric oxide, a chemical that makes the blood vessels in the penis expand. When this happens, the vessels fill with blood, which itself seems to block the blood leaving the penis, and, well, we all know what that means.

Interestingly, this is a bit different from what happens with Viagra. With Viagra, and other types of medications used today, the chemical process involved can only maintain an existing erection, while the spider venom actually stimulates an erection in the first place. While this might seem like a minor difference, there are people who don't respond to the pills on the market today. What Dr. Leite's research might lead to ultimately is a new option for those who need help, but can't find satisfaction with existing products.

In the meantime, no one's recommending you keep spiders in your bedroom.

TERMITE HEAD-BANGERS

Communication is hard for the lowly termite. Birds have their voices to sing to one another. Some animals use flashes of colour or big visual displays to get their message across. Still others use chemical cues to notify their neighbours of what's going on. But when trouble looms on the horizon for a termite, when the nest is threatened and it needs to call for aid, all it has is its head. And the way a termite uses its head would leave even the thickest of skulls shaking. Dr. Tom Fink, from the National Center for Physical Acoustics at the University of Mississippi, has figured out just how these wood-chewing insects sound the alert.

What they're doing is hitting their heads against the walls of their nests really, really hard. Dr. Fink looked at the two types of termite, the Formosa termite (*Coptotermes formosanus*), which he found bashes its head with forty times the force of gravity, and the native termite (*Reticulitermes flavipes*), which hits even harder, with seventy times the force of gravity.

He discovered that this hitting is quite carefully planned out. As in real estate, it's all about location, location, location. They'll seek out spots in their nest, known as a carton nest, where the wall is nice and flat. Since the wall is made of well-chewed pieces of wood, saliva, and mud, it's quite resonant, and the sound can travel through the wall, into the soil, and get picked up as vibrations by other termites. The other termites feel the vibrations through their legs, which send a signal to the brain, alerting them to the danger.

Just as interesting as the fact that they bang their heads is how they do it. It sounds, to a human, quite painful. The termite, first of all, raises its head up. Because of the termite's anatomy, this also brings up the first section of its thorax. Between the two thorax sections, there's a hinge. So the termite swings its head and thorax back, and then smashes its head against the wall. And what part of the head makes contact? "Actually," says Dr. Fink, "it would be like slamming your face, your nose, and your eyeballs, and everything straight into a table." Which is enough to give most people a headache.

You would think this might seriously damage the diminutive insect. But Dr. Fink says he's seen termites banging their heads for as long as two and a half minutes, which suggests that they aren't hurting themselves. While no one's looked in detail, they probably have plenty of fluid in the head to cushion the blow, as well as muscles and structural adaptations to make it easy to withstand the stress.

There's a practical side to understanding what's going on here. Termites are a significant problem in the southern United States, particularly the Formosa termites. These insects can infest trees and eat the heartwood right out, leaving the outside standing. These trees look perfectly healthy, but when a storm comes along, down they fall, damaging buildings around them. To figure out whether a tree contains termites or not, you stick a steel rod into the ground among the roots. "What happens," says Dr. Fink, "is that in the base of the tree is the carton nest and in the carton nest there are thousands and thousands of burrows. The termites perceive the sound of that rod going into the soil, feel the soil movement, and the soldiers will start head-banging."

On a big tree, which might contain hundreds of millions of termites, the head-banging can last for several hours. Even in a less infested tree, the termites will bang for forty-five minutes or more. Interestingly, not all the soldier ants will head-bang at once. It seems that a segment of the population head-bangs, while the rest get down to protecting the nest. And once worker termites hear the banging, they'll head towards the noise to fix whatever damage the soldier is telling them about.

It turns out that listening for the sound of banging may be one of the most effective ways of telling whether you've got termites in your trees. In a survey done by Dr. Fink and his team, they were able to find 97 per cent of the trees that had termites in them when they used acoustic tools, but only about half when they inspected the trees visually. If you've got rid of the termites in your house, it'd be worth taking a listen to any nearby trees, because many reinfestations of homes come from the termites that survive in the trees.

So the next time you're out in the woods, stick your ear against a tree. You just might hear the knocking of a head-banging termite.

Army Ants Lie Down on the Job

Napoleon is believed to have said that an army marches on its stomach. Well, he might have been right about humans (and of course he was really talking about food), but that's certainly not the case for army ants. No, a phalanx of army ants marches on another body part: the backs of its sisters. It seems army ants in South and Central America (*Eciton burchelli*) literally lay themselves down for the good of the colony and let their relatives walk all over them. Professor Nigel Franks, an animal behaviourist at the University of Bristol in England, has worked out why.

This particular species of ant lives in the rain forests of Panama. Like other ants, they live in colonies consisting of a single queen, female worker ants with various specialized roles, and larvae in various stages of development. The ants themselves aren't particularly spectacular, but the number of them in a single colony is astonishing, up to half a million individuals. When you consider that these ants are all related to each other, that's quite the family get-together.

Army ants go out in groups when they're raiding for food. In the case of this species, there can be as many as 200,000 ants travelling together on a single raid. "As a result," says Professor Franks, "they deplete their prey so much that they have to go and find prey elsewhere." If you're a scorpion, cockroach, or spider in the leaf litter where these ants are hunting, you're basically a guaranteed goner. The sight is impressive. Says Professor Franks, "They're fantastic. They look like the silhouette of a tree laid down. So you've got, at the swarm front, a very dense, huge group of ants. They're so dense that they actually colour the rain forest floor black with their bodies." Behind this front, and going back to the nest, the ants form a thick trail, which gets thinner and thinner towards the entrance.

In a given day, an army of these ants needs to bring about 30,000 prey items home, which creates a huge haulage problem. So some of these ants, referred to as a caste, specialize in carrying heavy objects. What attracted Professor Franks's attention was that some ants were coming back to the nest empty-handed. That looked like a waste of effort, except that they were, in fact, doing something. Professor Franks says, "What they do is essentially block holes in the forest floor with their own bodies and allow other workers – the rest of the colony, if you will – to run over the top of them."

That really understates how intricate this hole-blocking manoeuvre is. Army ants come in plenty of different sizes, from very small to quite large. And when they're using themselves to fill holes, they'll self-select to fit in the perfect hole. Professor Franks says, "If a small ant finds a small hole, it will plug it. But if a big ant comes across a small hole, it will ignore it. If a big ant comes across a big hole, it will plug it. So they match their own body size. But if the hole is too big for any one ant to plug it, one ant will attempt to do so and another ant will join in and help. You can have three ants, sort of spread-eagled over a large pothole, and the rest of the colony running over the tops of them." Just as human engineers create bridges to get over gaps in our roads, these ants are doing the same thing with their own bodies.

The good news for the ants here is that it doesn't hurt them. "When traffic ceases," says Professor Franks, "these literally downtrodden ants will extricate themselves from the hole, pick themselves up, and run off back home with the rest of their nest mates."

Why do they do this? Back in the lab, Professor Franks and his colleagues set up planks for the ants to walk across, complete with holes in a variety of sizes. By measuring the ants' speed as they crossed the plank, they were able to figure out that the use of a few ants to fill the holes dramatically increased the overall speed at which the colony could get food to the nest. To the group as a whole, then, there really isn't a cost. The needs of the many are outweighing the needs of the few, to the colony's eventual benefit. The colonies, you'll remember, are all female, so perhaps this is a gender thing. It certainly is the ultimate in division of labour, a strategy that ants are perhaps one of the best groups in the world at accomplishing. The question is this: is there anything we can learn about our own endeavours from them?

Loony Tunes

For many Canadians, summer means lounging in a deck chair, on a dock, at dusk, listening to the haunting sound of a loon call from across the lake. It's an idyllic summer scene, but have you ever thought about what that loon (*Gavia immer*) is saying to the world? Maybe it's saying, "Ah, what a beautiful evening," or "Come on over, ladies." If only we had a translator, such as Dr. Charles Walcott. As a professor at Cornell University, he's been studying loon communication, and he has found something that may cause ornithologists to change their tune.

Loons make certain types of calls that seem to have specific meanings. There's the wail, a series of long notes that seems to mean, "Come here." It's used to attract groups or pairs of loons together on a lake, and it's probably the cry you've heard from the dock. Then they have the warning cry, called the tremolo, which some people describe as insane laughter. These notes are much shorter, and you'll hear them if you get too close to a loon out on the lake. Then there's a hoot, a very short, soft call, but you're not likely to hear it unless you're close to a mother loon and her chicks.

Those calls are all made by both males and females. There's one more call, though, and that's the yodel. As Dr. Walcott says, "Only the male produces the yodel, and it's more analogous really to a songbird's song, the robin on top of the tree that sings cheerio." Each male has his own yodel, and it's subtly different from the yodel of other males around. The question in Dr. Walcott's mind was, how stable is that yodel over time? Male loons are very territorial, so one possible way of counting them, rather than catching and banding them, would be to listen for their calls. If the male's unique call stayed the same throughout his life, then you'd always be able to find him again. This could help with conservation, and be a much less invasive way of tracking birds than physically handling them.

At first, Dr. Walcott's research went very well. He travelled around his field study area at the Seney National Wildlife Refuge in Michigan's Upper Peninsula, and around Rhinelander, Wisconsin, identifying males. Sure enough, he and his team soon learned to tell the males apart, as each had a unique call. But then, says Dr. Walcott, "a loon changed its territory. And then, to my horror, it seemed to change its vocalization. This is something it should not have been able to do."

These changes were quite subtle and not easily detected by ear. But when Dr. Walcott compared his recordings of the call of this male, which he had made three years in a row (once before the bird moved, and then twice after he changed location), he could see on the sonogram a definite shift in the sound of his yodel. This was a huge surprise. In most bird species, the males are either born with the call they'll have all their life, as chickens are, or they learn their call when they're young. No one's recorded a call actually changing in adult male birds before.

The big change in the loon's life had been its move from one lake to another. At the second lake, it had found a mate, but had had to challenge another male to win her attention. The loser loon had left the lake in search of a more faithful mate.

Why would the loon alter its yodel? At first Dr. Walcott thought it might be because the male had a new female, as if, Dr. Walcott says, "the new female has said, 'George, I cannot stand that yodel and you'd better change it.'" But when Dr. Walcott tested this by looking at cases where females had moved to new lakes, he found no evidence to support the hypothesis. So instead, he looked at what the call of the male who'd been displaced had sounded like. "In every case, the loon changed its vocalization to be as different as possible from the previous resident," Dr. Walcott says. "For some reason, it's important for the loon to say, 'I'm a new boy on the block,' as it were."

This sets up three new facts about loon communication. First, says Dr. Walcott, they can change their vocalizations. Second, they make that change when they change territory. And third, the change is intentional and done to make them sound like a new resident. All this shows a great deal more sophistication than anyone had given loons credit for before.

It also creates a problem. It's not really possible to use caller identification for loons in long-term studies, since the males might move and mess up all the data. And it raises the question, what are they talking about, anyway?

Electric Fish

You know what they say: if you can't beat 'em, sabotage 'em. Okay, we've taken some liberties there, but there are plenty of examples of sabotage in the human world. Like a jealous co-worker who "forgets" to pass along that important phone message to a rival at work. Or a disgruntled employee who shuts down the office e-mail to get back at his boss. What's true for humans is apparently also true for the brown ghost knifefish (*Apteronotus leptorhynchus*). Except that, according to Dr. Sara Tallarovic, a researcher at the University of the Incarnate Word in San Antonio, Texas, their sabotage is truly shocking – they actually use electricity.

The brown ghost knifefish are, in Dr. Tallarovic's words, "not a very charismatic fish." They grow to about twenty centimetres (eight inches) long, and are an olive-brown colour. The males, says Dr. Tallarovic, "have almost a horse-like face and a long, knife-shaped body, with a long, extended anal fin and a very long, narrow, tapered tail." They are indigenous to South America, particularly the rivers of Colombia and Venezuela.

What's more important than their looks is that these are what are known as "weakly electric fish." They are cousins to the electric eel, but their use of electricity is different. Electric eels use shocks – pulses of reasonably strong electrical current – to stun their prey. That's not what these weakly electric fish are doing. Instead, they generate a very low-level, constant electric field around themselves. You'd never feel it if you stuck your hand in the water and touched one, but it's there; it can be picked up with electrodes.

These fish use this low-level, constant field to navigate through the water. The way it works is that different objects in the environment around the fish will conduct electricity in different ways. So, when a knifefish gets close to, say, a plant, it will detect a change in its electric field, thanks to the array of sensors it has on its body, and navigate around the object. The same is true of prey. Any nearby prey will cause a change in the electric field around the knifefish, and the knifefish can tell where it is.

Here's the interesting thing. All the knifefish produce an electric field, but each one of them produces a field at a particular, and different, frequency. It's a bit like each fish singing a slightly different note, except the note is an electric field. They'll hold this one note for months at a time, although they can, for short periods, change the note. The frequency of their electrical discharge, Dr. Tallarovic says, is their way of communicating with each other. For instance, during courtship, the frequency can change by hundreds of hertz away from the baseline, like the pitch of a voice rising or falling. For those with an electrical bent, or who want to amaze people at their next cocktail party, the frequency of a female knifefish will range between 600 and 800 hertz, while a male's will be from 800 to 1,000 hertz.

What does this have to do with sabotage? Well, it's a game that the male knifefish plays. First, a word about knifefish dominance. Brown ghost knifefish males have a hierarchy: the dominant fish is a bully, with higher androgen (i.e., testosterone) levels. It turns out that the most dominant male in the group is also the knifefish with the highest-frequency electric field. In other words, if you introduced two male knifefish to the same tank, you could tell in advance which one was going to be the top fish, just by measuring the frequency of their electric fields. Higher frequency means more dominant.

But what actually happens if you put two knifefish together in a tank? That's what Dr. Tallarovic did, and she noticed a strange behaviour: "What I found when I started pairing fish, to see if one was dominant over the other, was that when they would fight, they did this bizarre jamming behaviour."

Here's how that played out. Two fish are placed in a tank together. They're both emitting their electric fields, one at a higher frequency than the other. The fish with the lower frequency – in other words, the subordinate – will try to displace the higher-ranked fish and rush in to attack it. As it rushes, it raises the frequency of its electric field, which has the effect of jamming the more dominant fish's field. Again, musical notes can help explain this. Sometimes when you listen to a constant note, like the hum of an airplane, you'll hear a second note, rising in pitch, getting closer and closer to the pitch of the hum. At some point, when they aren't quite one note, you'll hear a change in the first note. It'll seem to get louder then quieter, almost like a beat. It's a form of interference, which in music sounds terrible, but in these brown ghost knifefish has the effect of jamming their electric fields. Which is to say that they lose the ability to sense the environment around them for as long as they're being jammed.

It isn't just the fish that's being attacked whose field is jammed. The same is true of the other fish, the one that's changing its frequency. But that doesn't seem to be much of a problem, says Dr. Tallarovic. "I've seen fish on video actually begin the jamming signal at the time they're darting forward to make a biting attack. What's probably happening is that even though they're briefly jamming their own signals, they know where they're going, and they attack pretty quickly. Probably they're going without their navigational ability once they've already started an attack towards the other fish. I liken it to jamming radar. If you know where you're going, you can briefly do without your navigational system, as long as you're distorting your enemy's perception of where you are."

The fish might try to avoid attack by increasing its own frequency, but the attacker will match it step for step. It's a sneaky system, and one the researchers hadn't expected to detect. It's a fish-eat-fish world down there, and what took humans until the twentieth century to develop in machines, these animals have been doing with their bodies for thousands of years. Yet another example of nature beating us to the punch.

CHIMP PLANS FOR PITCHING

This is a story about Santino, the chimpanzee. Santino lives at the Furuvik Zoo in Sweden on an island surrounded by a moat filled with water. Visitors gather on the far side of the moat to watch him. But Santino isn't a fan of visitors. He wanted to get rid of all those curious humans and their noisy children watching and pointing at him. To do this, he needed a plan. And, with a little thought, he devised one. The only problem is that Santino was not supposed to be able to do this. By coming up with a way to take out his frustration on zoo visitors, Santino demonstrated a capacity that surprised many behavioural scientists. Dr. Mathias Osvath, from Lund University in Sweden, brought the story of Santino to the world.

The tale begins with a behaviour that for male chimps is no surprise. When a male chimp wants to maintain his dominance, he'll demonstrate or show off. Usually this behaviour involves chasing other males, yelling, chest-beating, and strutting his stuff. But that doesn't work when there's a pane of Plexiglas between you and the creature you're trying to intimidate. All around the world, chimps in zoos have developed another strategy. They throw stuff. Food, rocks, feces, it all gets chucked at the windows. Which, as Dr. Osvath says, "has a very drastic effect."

This is what Santino was doing. But one day, when Santino was demonstrating at the guests, his keepers noticed a shower of stones coming over the heads of the crowd. When the keepers went into the enclosure that night, they found distributed around Santino's area, on the side close to where visitors appear, caches of stones. As this was only a quarter of the area of the island where Santino lives, it looked as if he knew where best to place the stones to throw at visitors.

Not really believing what she saw, the head keeper went in the next day to watch Santino before the zoo opened for business. "And then she observed the chimp reaching into the water," says Dr. Osvath, "and he was picking up stones from the water, putting them in these nice piles, and he was doing this completely calmly, almost sleepily, because it's early in the morning and chimps are like humans, a bit sleepy. And then later on, before noon, when visitors started arriving, he did his usual display. And in this display, he used the ammunition that he'd put out."

The keepers were astounded. Santino wasn't being provoked when he set out his stones. There were no people around to throw the stones at. He had to anticipate that later in the day there would be crowds, and that he'd need the stones if he wanted to throw them and intimidate people. This is something only humans are supposed to be able to do. Caching isn't that unusual in the natural world. Lots of birds and small mammals cache food for the winter. The difference is, they're probably obeying an instinctual response to the change in season and the ripening of the kinds of foods they harvest. But there's no instinctual basis for Santino's behaviour. He was acting in a logical – some might say rational – way.

The story doesn't end there. After seeing Santino's rock collection, the keepers removed all stones from the moat. That didn't stop the chimp. He just developed a new strategy. "He realized," says Dr. Osvath, "that in the middle of these islands, there are some rock structures made from concrete. And when the water seeps in through these micro-cracks in the summer and the autumn, they will freeze later in the winter. So they detach the outer layer of the concrete. But you cannot see this. You can only hear this by knocking on the concrete structure. When you hear a hollow sound, you know it's detached. So this was what he started to do." He tapped at the rocks, and when he heard the telltale hollow noise, he would bash the loose concrete off and break it into pieces of a size he could throw. In other words, Santino started making tools to throw.

The cold war goes on today. Santino makes rock projectiles to throw. The keepers take them away. So he makes more.

This is incredibly sophisticated cognitive behaviour. Santino is predicting the future, and his role in it, which we humans have always assumed was something only we can do.

But for Dr. Osvath, Santino's forethought wasn't a surprise. After all, he says, chimps live in a complex environment in the wild. They need to use tools to crack open nuts, fish out insects from cracks. They have to hunt, find ripe fruits, fight other groups of chimps. All this likely requires planning. The problem has been witnessing whether chimps in the wild are thinking ahead. Santino has given us the clearest example that forethought really happens.

There's a sad footnote to the story: the zookeepers were unable to control Santino's aggression, and they decided to castrate him, in the hope that his hormone levels would decrease, which would make him less prone to throw stones. He's apparently already getting fatter and likes to play much more than before.

CROCODILES GO HOME

What do crocodiles and boomerangs have in common? Well, they're both native to Australia. But nature has another answer to the question. According to Dr. Craig Franklin, a biologist at the University of Queensland in Australia, the crocodile (*Crocodylus porosus*) has a remarkable homing ability. In fact, take a crocodile away from home and it will return, just like a well-thrown boomerang.

Australia has two kinds of crocodiles: the freshwater crocodile (*Crocodylus johnstoni*), which lives mostly in small freshwater holes and is relatively small, and the estuarine crocodile (*Crocodylus porosus*), the one Dr. Franklin has studied. It's the world's largest living reptile, and it can grow up to seven metres (twenty-three feet) long. It's the one that most people picture when they think of an Australian crocodile.

Partly because of their size, and the problems that can happen when they come in contact with humans, the roaming patterns of these estuarine animals have been studied for several years. Dr. Franklin was interested in whether these animals could be moved away from where humans and livestock live, without harming them. As it stands right now, when these big crocs start wandering into human territory, they can't be chased off. The only solution is to capture them and put them in zoos. When that happens, their genetic stock is lost from the wild population, so Dr. Franklin wants to know if there's a better way of dealing with the problem of marauding crocodiles.

He and his team captured three big male crocs and moved them away from their homes. All of them originally came from an area of Queensland called Cape York Peninsula. Two of them were taken out of their rivers and moved by helicopter about fifty kilometres (roughly thirty miles) up the coast. The third was taken across a mountain range and released on the opposite side of the huge peninsula where he'd lived. The direct line over the mountains from his river home was about 130 kilometres (about 80 miles), which is far enough. The route by water was closer to 400 kilometres (250 miles).

Here's the boomerang part. All three of the animals made their way home. The pattern of movement was pretty much the same for all of them. They hung around the new place for a few weeks, and then made a beeline back to their original habitat. That was fairly straightforward for the two that were moved along the coast, but incredible for the one that had been moved across the mountains. "On December 4," says Dr. Franklin, "it decided it was ready to go home. It swam northwards, circumnavigated the tip of Cape York Peninsula, then swam southwards and straight back into its river system. The incredible thing was that it covered over four hundred kilometres [250 miles] in twenty days, and clocked up, one day in particular, over thirty kilometres [19 miles]." This amazed the researchers. While it was suspected that crocodiles preferred to stay in one place, no one had expected this level of homing ability.

You might be wondering how the research team was able to track these animals in such detail. Two words: crocodile wrestling. A satellite transmitter had to be attached to each animal's back. And no anaesthetic was used to sedate these monsters. Instead, they were wrestled down and held while the transmitter was attached. Dr. Franklin says, "It's almost like giving the crocodile a big hug. It's essentially using body weight to hold the animal down. But in saying that, if the animal wants to move, you move with it, and you just get to hold on tight. And hopefully things go well. And, to date, they have."

You don't do this alone, of course. It takes a team of six to tackle a 4.5-metre (15-foot) crocodile. "We always put the biggest guy at the front, naturally," says Dr. Franklin, "because even though the jaws are secured, you have to jump very quickly, and you jump in concert. So we have it like dominoes. The first person jumps on the head, and then the next person jumps on top of the person on the head, and then one on the neck, and one of the forelegs, body, hind legs, and tail." Luckily for them, the team included the late Steve Irwin (a.k.a. "the Crocodile Hunter"), who taught them how to do it. It is not something for amateurs to try!

This opened new doors for crocodile research. For all their size, crocs are shy creatures, and this was the first time anyone had been able to track them so well. It turns out that they have innate abilities no one ever suspected. Where do those abilities come from? Dr. Franklin says, "The interesting thing about crocodiles is that they're very closely related to the birds. In fact, they fall within the same lineage as the birds. And so crocodiles are more closely related to birds than they are to all other reptiles." We know that birds have great homing abilities, so maybe it's not so strange that crocodiles do, too. They may be using magnetic information in the same way birds do, or they may be able to detect the specific scent of their home river, using their very keen sense of smell. The question of how they know which way home lies remains to be figured out, and with it the answer to the original question: whether it's possible to relocate a crocodile and not have it return home.

In the meantime, don't confuse a boomerang with a crocodile. Try to throw a crocodile and you might get a nasty nip.

Alligator Air Bags

Certain animals are more challenging to study than others. The alligator, for instance. One reason we don't know as much about this animal as we do about finches, for example, is that the toothy reptile isn't exactly a friendly research subject. Close examination of an alligator's anatomy is more likely to lead to the emergency room than the lab bench. But not for Dr. T.J. (Todd) Uriona, a researcher at the University of Utah. He's comfortable getting close to these creatures, and that's put him in a unique position to find out just what makes them tick. In the course of his research, he's discovered what may be the secret to an alligator's underwater stealth and agility.

The secret is a set of muscles, associated with their lungs, that's not found in other animals. As Dr. Uriona says, "Prior to our study, most people just assumed these muscles were to help them breathe. However, we were interested to see if maybe they could use these muscles to pull their lungs around inside their body, and then, in doing that, change their aspect – to be able to tip their head down and get their back end up, or tip from right side to left side just by moving air around inside their body."

Yes, that's right, we're talking about using muscles to move their lungs around in order to change the way they lie in the water.

Picture an alligator submerged in the water. The weight of the alligator holds it down under the surface. But inside the animal is a bag of air, its lungs. When the bag is perfectly balanced in the centre of its body, the alligator will float flat in the water. When it moves that air to the front, that end becomes more buoyant and its snout tilts up. When it moves the air back, the hindquarters rise. When it moves the air left or right, then the animal rolls. It's a way of steering that doesn't require the alligator to thrash around. If it's small and trying to avoid a predator, or simply on the hunt, moving by forcing the air around inside its body is a great way for the alligator to disguise its activity.

There are several muscles involved in this process. The first is one that humans are familiar with: the diaphragm. In us, it separates our heart and lungs from the rest of our organs, and when it contracts, it pulls air into our lungs. The alligator's works differently. "It attaches to their hips and then comes up, runs the length of the body, and attaches to the liver," says Dr. Uriona. "So when it contracts, it will pull the liver down towards the hips. And the liver is attached to the lungs. So it's actually pulling the lungs back in the body cavity towards the hips."

That's not all. Like humans, alligators use muscles in their rib cages to help expand their lungs when they're breathing, but then they use the same muscles to move the lungs around underwater. And then, says Dr. Uriona, "They have another group that's associated with their pelvis that allows them to pouch out their belly, and in doing so, increase the volume that the lungs can fill." The bigger the bag made by the lungs, the more useful it will be for manoeuvring under water.

Just how do you go about measuring the movement of lungs in an alligator? The answer is, of course, carefully. And Dr. Uriona isn't foolhardy. For his research, he used juvenile alligators less than a metre (a yard) long, in which he implanted electrodes to measure muscle movement. Then it was a case of long hours watching them swim in a small tank, observing what was happening. And sure enough, they were using these muscles to move up and down and side to side.

It's an elegant system, and if you've ever seen pictures of an alligator breaking the surface of the water, or slipping back below the surface, with scarcely a ripple, you know how eerie their movement is. And they're probably not alone in using their lungs this way. Some frogs and turtles have similar muscles around their lungs. Dr. Uriona thinks it's possible this kind of control system has evolved many times in aquatic animals. Which makes you wonder, why didn't he start his research with a less ferocious animal?

Bird Doping

In recent years, the use of performance-enhancing drugs in sports has frequently made the headlines. Scandals involving the use of growth hormones, steroids, and amphetamines have hit every sport, from sprinting to baseball, from cycling to darts. But it's getting worse. Now, it's showing up in the animal kingdom. At least one species of bird has been found to ingest performance-enhancing substances to help them endure the marathon of migration. The culprit is the semi-palmated sandpiper (*Calidris pusilla*). Dr. Jean-Michel Weber, a biologist at the University of Ottawa, uncovered this bird's doping strategy while studying its migratory habits.

The big issue for these birds is the huge distance they fly when they migrate. They start out in the Canadian Arctic and end their journey in South America, as far south as Uruguay, after travelling more than 4,500 kilometres (2,800 miles). On the way down, they travel fairly slowly until they reach the mud flats in the Bay of Fundy, between New Brunswick and Nova Scotia. But after that, they make the remaining journey in just three days. Dr. Weber says, "This is where the scandal takes place, apparently."

While in the Bay of Fundy, the sandpiper eats, and eats, and eats. In the two weeks these birds spend on the mud flats, they'll double their mass, from twenty grams to forty grams (three quarters of an ounce to an ounce and a half). These birds are bulking up on mud shrimp. Dr. Weber describes it as "gorging." And, he says, "what is really interesting here is that not only does the bird get a lot of energy out of the lipids [the fats within the shrimp], it's also getting something more. What it's getting is a substance in one of the lipids which really improves its performance."

Here's where the doping part comes in. The compound is one you've probably heard of: the omega-3 fatty acid. Omega-3 fatty acids are good for human health and nowadays are added to some foods, but in this case, we need to look at what they do for a bird.

To understand what's going on, we need to take a quick diversion into biochemistry. There are two features of omega-3 fatty acids that could be important here. First, when they're present in the membrane of a cell, they make that membrane more flexible. In the case of bird cells, it means that other fats, which are the main source of a bird's energy when it's flying, can get into the cells more easily. That's getting the fuel to where it's needed more efficiently. Second, the omega-3 fatty acids stimulate the cell to produce more mitochondria. And mitochondria are the part of the cell that breaks down the fat and converts it to energy. In other words, adding a lot of omega-3 fatty acids to a bird's diet helps it get fuel into the cells faster, and then burn that fuel more efficiently.

This extra boost from the shrimp is crucial to the sandpipers' survival. It's a very tough flight down, and the birds arrive extremely tired, and much lighter. They even use up part of their muscles near the end of the flight. (Not that they need those muscles to be as large any more. Once the fat is gone – or the "suitcases," as Dr. Weber puts it – then the muscles can shrink.) It's not clear how many birds survive the migration, but it's certainly not all of them. And recently, there has been a decline in both the shrimp and sandpiper populations, which, based on this evidence, presents a tempting correlation to explore. Dr. Weber thinks that if the shrimp populations disappear, the sandpiper won't be far behind.

Now, whether gorging on mud shrimp will become the next body-building supplement craze has yet to be seen, but upping our omega-3 fatty acid intake can only be a good thing.

Scuba Bug

There are few experiences that can compare with scuba diving. But, for humans, the experience has one big limitation. We can explore the undersea world, but only until we run out of air. Good dives always seem too short, since we have to surface before the air tank runs dry. That's not the case if you're a scuba bug – one of the several species of diving insects that have their own underwater breathing apparatus. These insects can stay underwater indefinitely; they don't have to come up to the surface to replenish their air. Dr. Morris Flynn, an engineer from the University of Alberta in Edmonton, has worked out what allows these insects to stay perpetually submerged.

There are hundreds of different insect species that can breathe underwater, but Dr. Flynn did most of his work on a group called back swimmers, which are native to New England. They're called that because, no surprise, they swim on their backs, which is probably the best way to identify them. Most of them are less than two centimetres (three quarters of an inch) long, and some of them will attack humans, although their bites, while painful, are harmless. If you've been near a lake or river, you've probably seen them. They're very common critters, but next time, look at them a little more closely, and see if you can spot their scuba gear.

There are two strategies diving insects use. One is similar to the tank of compressed air that humans use. When these insects dive, they take with them a bubble of air. That air contains oxygen, so as they dive, the insects will take in the oxygen from the bubble. When the oxygen is all used up, they head back to the surface to get a new bubble and dive down again.

The second strategy is much more interesting, and something humans have yet to try. Like the first group of insects, these bugs carry a bubble with them when they dive. The difference is that their bubbles don't act as a scuba tank; instead, they are like a lung. Here's the explanation. The role of lungs is to exchange gases. Air flows in, and oxygen passes from the lung into the bloodstream, while carbon dioxide goes in the other direction. The same thing happens with these bubbles. The bubble sits on the outside of the insect, which takes oxygen from the bubble into its body. That makes the concentration of oxygen in the bubble lower than in the surrounding water, so oxygen passes from the water into the bubble. Carbon dioxide goes in the opposite direction. With these gases passing between the bubble and the water, as long as the insect's metabolism isn't too high (i.e., it doesn't use too much oxygen or get rid of too much carbon dioxide), then it can stay underwater indefinitely.

Here's the issue, though, and the question that Dr. Flynn wanted to figure out. How does the bubble stay stuck to the insect? The first part of the answer is simple. Dr. Flynn says, "What the insect does is to hold on to the bubble with a series of very long, waxy hairs." When the insect dives below the surface, air bubbles are stuck among these hairs. As well, the insect's outer shell is so water-repellent that, says Dr. Flynn, "instead of having those bubbles getting dislodged, more often than not they'll actually stick onto the insect's body."

This raises another problem. If the insect is covered in bubbles, why doesn't it rise up to the surface like a cork? In a lot of cases, the bubbles are minuscule, which helps, but there are some species that do face a problem with buoyancy. Their solution? To cling to the bottom, or secure themselves to vegetation.

For the rest, those that can float underwater with bubbles on their surface, there's a different set of concerns. Dr. Flynn says, "If there are too many hairs, then the bubble gets smothered, and there's not enough surface area for an exchange of gases – oxygen coming in and carbon dioxide going out. On the other hand, if there are too few hairs, then it's very easy for the bug to lose the bubble." Not only that, but as the bug dives deeper, the water pressure changes, which alters the shape of the bubbles and, along with it, the shape of the hairs. Dr. Flynn plugged all these different factors into his computer models to try to understand what's going on when the insects dive.

What he found was that the pattern of hairs on an insect needs to change, depending on how they want to dive. "We found," says Dr. Flynn, "that insects that want to dive quite deep in the water column require a very tightly packed hair lattice, so that the space between adjacent hairs is relatively small." It's a matter of physics. In deeper water, the higher pressure pushes on the bubbles and causes them to flatten out. The tightly packed hairs can hold on to these flat bubbles, which means there's enough surface area on the bubble pressing against the insect's body to allow the gas exchange. In shallow water, on the other hand, there's less pressure, which makes the bubbles rounder, and the bugs swimming at this depth don't need as many hairs to hold on to them. "What's surprising," says Dr. Flynn, "is that our model predicts that insects that are adept at diving to relatively large depths, on the scale of an insect, may have trouble surviving at very shallow depths." Different strokes for different folks.

As interesting as this discovery is, unfortunately it won't help humans dispense with scuba gear any time soon. "Humans are big," Dr. Flynn says, "and being warm-blooded, our respiratory demands are simply too big to make this a feasible technology. On the scale of a human, you'd need a bubble that's the size of a small office in order to have enough oxygen coming in to meet a human's oxygen demands." We'll just have to stick with tanks and masks, and limited trips into the ocean.

Head-Under-Heels

The next time you want a real physical challenge, try this (or perhaps not). First, find a set of monkey bars. Then, get a good running start at them and try to grab hold. Not with your hands. With your feet. Upside down. Seem a bit daunting? Impossible is more like it. But it's what bats have to do when they come in for a landing. Despite the challenge, they do a fantastic job. Exactly how they do it is a matter that eluded researchers until Dr. Daniel Riskin, a Canadian biologist at Brown University in Rhode Island, looked into the question. He has figured out exactly how they perform their aerial acrobatics.

There's a good reason to ask this question. Bats have long, skinny arms and legs that are all bound up in their wings. But flying isn't all that bats do; they also have to rest. How do they avoid crushing those long, skinny limbs when they come in for a landing?

Or, as Dr. Riskin puts it, "I was curious about what happens when a bat lands, and whether or not those legs get squished."

Different bats face different challenges when they come in for a landing. Some, such as the dog-faced fruit bat, a very small flying fox bat, roost in foliage. When coming in to land, these bats have to be able to grab on to a branch or leaf. Others, for instance the leaf-nosed fruit bats from the New World tropics, roost in caves and grab on to the surface of the rock. These are quite different circumstances, and could mean different strategies for landing.

One of the problems faced in studying this is that bats don't hang around when they want to come in to land. In fact, when Dr. Riskin started studying this at Le Biodôme in Montreal, "the bats just weren't co-operating, and I wasn't able to quite see what was going on." So he moved his work to Brown University and built a much better system. "I built a bigger force plate and better cameras – three of them," he says. A force plate is an instrument for measuring the force of any pressure applied to them. Bathroom scales are a simple force plate. Dr. Riskin covered his in mesh, so that the bats had something to grab on to. "The way we set up the experiment was to put the big force plate on the ceiling of a room, and then to set up plastic sheeting so that the rest of the ceiling sloped down from it, so that the highest point in the room was the force plate. The bats would naturally go for the highest point they could land. They would fly in and would land on the force plate. And as they did that, I had three high-speed cameras filming them at a thousand frames a second."

Just as you'd expect, the two kinds of bats, those that land in trees and those that live in caves, came in to roost in different ways. One came in with a two-point landing, the other a four-point landing. We'll look at the four-point landing first.

The bats that land in leaves are the ones executing a four-point landing. As the bat approaches the ceiling, its nose points up, higher and higher, until it's pointing straight at the ceiling. Then the bat flips onto its back, with its belly pointing towards the ceiling. At that point it makes contact with all four feet – both hands and feet, so to speak. That's the four-point landing.

Then there are the two-point landers. These are the bats that typically roost in caves. It starts out much the same. The bat's nose pitches up and up, but, Dr. Riskin says, "when they got to almost vertical, they would suddenly do almost a cartwheel, like a side flip. They would bring the legs up the side of the body and grab the ceiling that way." When these bats grabbed it was with their hind limbs, or feet, only. In aeronautical terms they had pitch (the nose going up), and roll and yaw (the sideways flip), allowing them to land upside down.

Whether it's the four-point or the two-point landing, this is an incredibly difficult manoeuvre. First of all, the bat can't hit the ceiling too fast, or it's going to injure itself. And it can't fly upside down, even though it has to land that way. At some point, it stops flying and becomes a projectile. The trick is to be moving hard enough to land, but not so hard it will be squished. And here's another difference between the two groups of bats. The tree-roosters, with their four-point landings, were, Dr. Riskin says, "nailing the ceiling pretty hard, up to eleven times their own body weight." The cave-dwellers were much more ginger in their landings, usually hitting with less force than their own body weight.

This makes sense. Trees have give; cave walls don't. When the bat throws itself at a branch, it will take some of the force. But a cave wall has no give, so the bat has to land lightly or it's going to be in a lot of pain.

Not that this explains why there are two-point and four-point landings. It may be because the bats in caves are landing where they're going to stay and don't have to worry about moving again. The tree-dwellers release two of their grips a few moments after they've landed, so they hang from just their feet, but they've got the option of adjusting their position after they've made contact with the branch.

Who knew how hard it was for a bat to land and roost?

BEES' KNEES

Here's a bit of a mystery for you. According to the laws of aerodynamics, bees shouldn't be able to fly, but obviously they can and do. While much attention has been paid to how their wings keep them in the air, far less has been paid to any other factors that might be responsible for keeping them airborne. But Dr. Stacey Combes, an insect flight specialist at Harvard University, has been buzzing with enthusiasm as she studies these rotund little fliers, and may have found the secret to their stability.

Dr. Combes has been working with a group of bees called the orchid bees (Euglossini). These are tropical bees that pollinate orchids. Unlike the striped bees we usually see, these ones are iridescent colours, mostly green, but some are red and bright blue. And the males of this group are a touch obsessive. They will go to enormous lengths to capture new smells. In the wild, the males collect these scents primarily from orchids, and store the compounds in sacs on their enlarged back legs. Their obsession makes the males of these bees easy to train. As Dr. Combes puts it, "We can go into the forest with a bottle of something, like cinnamon oil or vanilla oil, and bees will just magically appear from the forest as soon as we open the bottle. They will do almost anything that we want them to do."

What she's got them to do is fly in a wind tunnel. She dabs a tiny amount of some interesting scent at one end of the tunnel, then releases a male bee at the other end. The bee starts to fly down, and then the wind starts. In most insects, that would be it – the bug would give up. But not the orchid bee. These males keep going as if their life depended on it. Dr. Combes increases the wind speed until the bee can't keep up. That's the maximum flight speed for the bee. Tops for this animal? About 7.25 metres (23.7 feet) per second, although most bees topped out around 7 metres (23 feet) per second. That's pretty fast for an insect. Although, since not many insect speed demons have been examined, it's hard to say how outstanding this feat of speed actually is.

The more interesting part is what happens to the bee's body as it speeds up. Normally, for something to move faster, it needs to be smoother, to reduce drag. That's what happens to a plane after takeoff, when the landing gear is pulled up. But these bees do exactly the opposite. Dr. Combes was surprised. "They have these huge hind legs," she says, "and the faster they go, they actually drop these hind legs down below their body, until they're dangling down way below their body. So they look really strange, and it was completely unexpected."

From the bee's perspective, there seems to be a good reason, though. It turns out these bees fly a little like helicopters. That is, their wings are fixed at a particular angle, which tilts them up and a little forward. When they want to speed up, they need to tip their wings farther forward and less up, which means pointing the front of their bodies down. To do this, they extend their hind legs to create drag. That drag pulls the legs backwards, and *voilà*, the head end tips forward.

That's not all that's going on. Dr. Combes noticed that the limit of speed for these bees came because eventually they lost stability. They'd start to roll to one side, re-stabilize, and then roll again. At some point, they'd roll too far, flip upside down, and fall to the ground. It turns out that the legs play a key role in keeping them stable. Not only are the legs creating drag, they're also creating lift. "They're actually like little airplane wings," says Dr. Combes. "That lift is going out to each side."

Everything is fine when the bee is flying in a straight line with its legs down. But when the bee makes a turn, the lift on one side will be more than the other side, and the bee will start to roll. So, the bee will draw in one leg or the other, to reduce its lift, and re-balance (remember, the other leg is still generating lift, so the bee will roll back in the opposite direction). Pretty advanced aeronautics for a tiny tropical insect!

Whether other insects use this technique isn't clear. Few researchers have studied anything about orchid bees other than their wings and how they feed. And the stabilizing mechanism of the orchid bee may not apply to other kinds of bees, as their legs are an odd shape – convex on one side, and concave on the other – which may be what generates the lift in the first place. Still, considering the interest these days in developing miniature flying robots, understanding how the bee's legs are used could prove very helpful. Keeping the tiny robots stable has been very difficult so far. Maybe adding a second control system, using the legs along with the beating wings, will solve the problem.

CLOWNFISH SEX

Most animals, regardless of their species, are either male or female by the time they hatch or are born, and that's the way they stay all their lives. No so the clownfish (*Amphiprion percula*). Not only is its sex indeterminate when it hatches, all the young members of this sea-dwelling fish species eventually become males; that is, until they become the most senior member of their group, and then they switch sex, to become the matriarch. Dr. Peter Buston, a biologist at the University of California, Santa Barbara, was the one to discover this gender-bending behaviour.

You probably know what a clownfish looks like. Think of *Finding Nemo,* the movie with the little orange-and-white fish that lived on a coral reef. That's the clownfish. They live in small groups among the tentacles of a sea anemone. It's a pretty good life for the fish. The anemone produces toxins that most fish have to avoid, but the clownfish has evolved an immunity to the stingers. The clownfish gets protection from predators, and the anemone gets different benefits, depending on where the two species are cohabiting.

Not that everything for the clownfish is easy. First of all, when a new clutch of clownfish eggs is laid, they stay under their father's care for about seven days. A couple of weeks after they hatch, the larvae settle onto the reef. The next step is critical: as they start to mature they head off in search of an anemone. According to Dr. Buston, "They locate suitable sea anemones using chemical cues, so they sort of sniff them out. And then they'll turn up at an anemone, and they'll try and enter, and if they're lucky, they'll get in. If they're unlucky, the resident clownfish there will stop them from entering."

The youngster has just one shot at finding a home. Those that don't make it into a colony don't live. Those that do survive, but there's a long road ahead of them. "If they do make it in," says Dr. Buston, "they're going to make it in at the bottom of a hierarchy that's already present in that anemone, and the hierarchy is a queue. There's anywhere from three to six fish in the anemone. They're all queuing, just the same way you or I would queue in the post office, except they're not queuing for stamps. They're queuing for access to breeding positions."

Here's the weird part. In each anemone, there's only one female. It's the largest of the fish, and the next smaller one is a male. The remaining fish are neither male nor female, but have both sex organs, and they are progressively smaller, down to the most recent addition to the anemone, which is the smallest. So far, nothing terribly unusual, except for this: when the female dies, the male gets larger, and, Dr. Buston says, "his testes shrivel up, and his ovaries grow. All the rest of the fish in that queue, up until this point, they were non-breeders. They weren't breeding at all. They were just biding their time. And then the largest of these non-breeders will start functioning as a male, and everybody will grow up one step in this size hierarchy."

Here's a situation where size really matters. And size is very carefully regulated in these colonies. In fact, a clownfish will never get larger than 81 per cent of the size of the next fish up in the hierarchy. How this is controlled isn't clear. It's not about the biggest fish getting the most access to food, since in experiments Dr. Buston was able to show that no matter how much food is present, each fish stays the same relative size. A more likely explanation, he says, is that the fish intuitively know that if they get too large, then the slightly bigger fish will kick them out of the anemone, which is guaranteed death. So it's a matter of self-control, a diet, if you will, to keep from becoming a threat to your neighbours.

This helps explain why some youngsters are successful in joining an anemone group and others are not. If they are small enough to not intimidate the lowest fish in the hierarchy, they're allowed to stay. Of course, it's just the beginning of a long wait, if they want to breed one day. Dr. Buston says, "These fish are known to be quite long-lived. There are records of fish living fifteen, twenty years in aquaria, but these are all tracking individuals once they've reached that matriarchal status. If you take into account they have to queue through the entire hierarchy, my estimates would put females at over thirty years old and maybe coming on for fifty years old." That's a long time to hang on, just on the off chance you'll get to breed at some point. But it's the only game in town. Leave, and you're not going to survive.

Brings a new twist to looking forward to growing up, though.

Worm Hibernation

The tiny nematode worm, properly known as *Caenorhabditis elegans*, is a creature of endless fascination to biologists. They've studied its genetics and neurology because, despite its simplicity, it shares enough characteristics with humans to provide important insights into our own biology. But in some ways, the nematode is very different. Take, for example, its remarkable adaptation called the dauer state. This is a sort of living death for nematodes. They don't eat, don't move, and barely age. While we do nothing like this, Dr. Richard Roy, a developmental biologist from McGill University, still thinks there might be some application for humans that we can achieve by studying the adaptation.

The dauer state is quite remarkable. The worm enters it very early in development, but only under specific circumstances. As Dr. Roy describes it, "They can somehow sum up how many animals are around them, and then make some sort of mathematical equation, if you will, based on the amount of nutrient resources that are also around them. And if that equation isn't favourable, then they'll opt out and they'll go into this dauer state." So, if the local soil is crowded, and there's not much food around, it's time to enter the dauer state. It's a form of suspended animation, and when it's over, the worm picks up exactly where it left off. According to Dr. Roy, "They'll come out, and they're absolutely fine. You'd never know that they actually went through this."

In order to enter the dauer state, the nematode does have to prepare. First, they seal themselves off completely from the outside environment. Nothing gets in past their cuticle, including food, and nothing gets out. Then they stop all cell division, which allows them to conserve energy. Finally, they readjust their metabolism so they start accumulating lipids (a type of fat), which they use as a source of energy while they're in the dauer state. Once all this is done, they can last as a dauer for up to six months. When you consider the entire lifespan of a nematode is two weeks, being able to wait around for half a year is pretty amazing. Of course, as Dr. Roy points out, "Unfortunately, when they come back, they have their final week of life and then they're kaput. They die."

It's their accumulation of lipids that most interests Dr. Roy. In his genetic studies of the worm, he found a single gene that controlled how fat was used. When the worm was getting ready to go into the dauer state, this gene was responsible for slowing down the burning of fat, allowing it to accumulate for the long wait. Then, once the dauer state was entered, this same gene turned up the conversion of fat to energy, but only enough to keep the dauer alive. In fact, in mutants, where this gene was turned up too far, the dauers burned out after just a few days. They were not able to regulate their fat use.

How does this help us? Well, Dr. Roy thinks this same pathway may be present in all animals that hibernate. There are plenty of species that have to be able to accumulate fat at some times and then burn it off in a controlled way later. Bears are a prime example. With his genetic study, Dr. Roy says he's figured out how to separate the accumulation and the burning of the fat. "In my opinion, if we really focus on things, we might be able to develop some kind of compound that would uncouple this and allow us to burn off fat specifically during a specific time, somehow."

To translate, he's looking at making a diet pill, one that would burns off excess weight in humans without our needing extra exercise. He's even looking at how to do this so that it targets only the "bad fat," the stuff that accumulates around our middles. And it might just be feasible. Dr. Roy says, "All of these genes are highly conserved, and it's probable their functions are also highly conserved." We've spent millions of years evolving to put on weight when there's an abundance of food, and then burning off the fat slowly during leaner times. This study of the nematode suggests that all the dieting fads of the last few decades are fighting against biology. In the nematode, when times are lean, it burns less fat than it does the rest of the time. If this is true for humans, when we diet, we start burning less fat than we would otherwise. Dr. Roy hopes to split these functions apart, so that we can turn up the burn when we want to lose weight.

Amazing what we can find out by looking under the microscope.

Four-Eyed Fish

One thing you can count on in nature is surprises. Just when you think you've figured out how things are done, nature finds a new way to do it – a new variation, or sometimes a whole new approach. The vertebrate eye is a great example. This amazing organ is shared by everything from fish to frogs, from birds to humans. It's a beautiful and complex structure which gathers light through a lens and focuses it on a retina, a structure assumed to be common to every vertebrate. Well, it's time to think again. Dr. Ron

Douglas, a visual scientist at City University in London, England, has found that a rarely studied fish, known as the spookfish (*Dolichopteryx longipes*), uses a whole new means of collecting light.

The fish is fairly new to science. It was found in 2008 in deep water between Samoa and New Zealand. Finding anything there is very difficult, partly because it is very dark at those depths and partly because there are very few animals there anyway. Collecting them is a very crude process. You put down a big net, about the size of a soccer goal, and drag it through the water very slowly, at about three kilometres per hour (less than two miles per hour). As Dr. Douglas says, "You have to be really quite stupid to be caught by that." After a few hours, you might have a small bucket of samples to look at that most scientists in the boat rush to grab. But Dr. Douglas had a different approach. "I came along about half an hour after everybody had taken what they wanted," he says, "and I noticed this rather odd-looking thing lying in the bottom of the bucket. I asked if anybody knew what it was, and nobody really had any idea. Somebody said, well it might be a damaged this, or it might be a damaged that. But, clearly, when I looked at it and picked it up, it was something that I had never seen before, that nobody else on the ship had ever seen before, either."

What he'd found was remarkable. "It was maybe about six inches [fifteen centimetres] long," he says. "It was white, with distinctive black dots along the side. And it was covered with this kind of weird, jelly-like goo. The really strange thing about it was that it looked like it had four eyes. And I knew that fish with four eyes just don't exist in nature."

Dr. Douglas's first question was, if this fish has four eyes, where is it looking? He took it back to the lab and photographed it from above, from below, and from the sides, to see if he could catch any flashes of light. Why look for flashes? Because, deep-water fish have mirrors behind their retinas designed to reflect light back over the sensing cells, to maximize the amount of light they can capture. Cats, dogs, and deer have them, too. That's why, on the highway, a deer's eyes light up at night when it looks at a passing car. This "eye shine," as it's called, is just us seeing the reflected light.

On first examination, this fish behaved like any other. There was eye shine when pictures were taken from above, showing that the fish was looking up. But then, much to Dr. Douglas's surprise, there was a second flash of eye shine from below. It looked as though the fish was looking up and down at the same time. This is, to say the least, unusual, and led to more examination.

Disappointingly, it turns out that this fish doesn't really have four eyes. It has the standard two, one on each side of the head. But each eye is divided into two parts, one for looking up and the other for looking down. The part that looks up is common to all vertebrates, with a lens that focuses light onto the retina. But the part that looks down hasn't been seen before in nature.

This animal doesn't focus light in its eye using a lens. Instead, it uses a mirror. We have ways of doing this ourselves, and some telescopes use this technology. A curved mirror will focus light down to a point. This can also be done with group of small mirrors, clustered in a curve. That's what the fish does. Inside this downward-pointing part of its eye, there is a set of small mirrors, made of tiny plates of crystals, that focuses light onto the retina.

Mirrors themselves aren't unusual in nature. Fish scales are an aquatic example that jumps to mind, and we've already mentioned the mirrors in the eyes of cats, dogs, and deer. What's different here is that these mirrors are acting like a lens and focusing light. It allows the fish to see below itself for flashes of bioluminescent light from other animals.

How this came to be is anyone's guess. But it's a very efficient way for a new structure to evolve. As you might guess, evolving a new lens would be a very complex, although not impossible, change. Mirrors, on the other hand, are much easier. They're already there, so the genes for making mirrors are in place in fish. It's simpler to move things around, which is what's happened here, than to evolve something brand new.

Once again, nature has thrown us a curveball. It will be interesting to see what we discover next that breaks the rules, as we understand them.

Cold, Dry Bugs

For many researchers, studying insects means travelling to the places where they're found. Jungles, especially, are environments teeming with bugs. But Dr. Richard Lee, from Miami University in Ohio, studies insects in an environment where you wouldn't think there are any. He works in the Antarctic, a frigid continent where almost nothing can survive on land. Penguins and seals spend part of the year there, but they spend much of their lives in the sea. Dr. Lee is studying an insect that spends its entire life on the land of the Antarctic, the farthest south that any insect has been found. It's a fly that has come up with some unique adaptations to survive the cold.

The insect is a wingless fly or midge (*Belgica antarctica*), or Antarctic fly. As Dr. Lee remarks, "Winglessness is a common phenomenon for insects that are found on islands or windy places, because if you get up in the air and are blown out to sea, you don't leave very many offspring." It's small, only three to four millimetres (a tenth to an eighth of an inch) long, and is dark in colour. As small as it is, it's the largest terrestrial animal on the Antarctic continent.

These flies are widely distributed over the continent. Dr. Lee has found them on offshore islands and on the western shore of the Antarctic Peninsula. They live around penguin rookeries and elephant seal wallows, and some even live on algae. The adults are out in January, which is the Antarctic spring. Within ten days they've laid their eggs, and many of the larvae survive long enough to overwinter just below the surface of the soil.

The climate isn't as extreme as you might think along the coastlines where they're found. For most of the summer the temperature is around the freezing mark, but it can rise as high as 25 degrees Celsius. In the winter, the air temperature will drop as low as −30 degrees Celsius, but where these insects hang out, the microclimate typically doesn't go much below −10 degrees. That's partly thanks to shelter and to snow cover, but largely because the Antarctic Ocean moderates the climate, the same way oceans do everywhere.

However, these insects have had to adapt to the constant cold and long periods of darkness. One of their adaptations is common to many Arctic and Antarctic species: they produce a form of antifreeze. In this case, two sugars, glucose and trehalose, fill up the cells in the fly's body and prevent it from freezing solid when the temperature drops. But if that were all the flies did, Dr. Lee wouldn't be that interested.

These flies have another adaptation, something that Dr. Lee calls "cryoprotective dehydration." He describes it as "a way in which the insect can dehydrate and significantly reduce its body water concentration and avoid freezing through the winter months. We were quite surprised to see this, because it had never been described before in a true insect."

The dehydration is quite severe. Dr. Lee says the flies lose between 50 and 70 per cent of their body mass by drying themselves out. It leaves them looking quite odd. "They look like little raisins," Dr. Lee says. "They look like they're dead for all the world. Then you add water and they plump up, and they wiggle away, and I think we can hear them laughing at us."

How exactly they accomplish this isn't clear. They do appear to be very permeable, so they dry out and rehydrate easily. If they're lying in the ground, and ice crystals form around them, the air itself is drying out, and water will evaporate from the fly. When that happens, the antifreeze compound in their cells becomes concentrated, so the insect never actually freezes. It's able to lower its own freezing point to match that of the soil around it. This allows the insect to drop its metabolism to almost nothing. Which is useful in winter, since there's not enough food to go around anyway. By turning itself off and waiting for warm weather, it's more likely to survive.

Understanding how this insect keeps itself alive has a practical purpose. While there are a few other animals out there that can survive freezing temperatures, we have yet to find one that gives a complete solution to a problem we want to solve. That's the issue of the cryopreservation of organs. If we could find a way to freeze organs for transportation or storage without their breaking down, this would help the field of transplantation. Right now, we have two problems. First, when we freeze tissue, ice crystals form, which break the cells and kill the organ. Second, we don't know how to keep an organ actually alive but dormant when it's frozen, so that we can bring it back to life. By understanding more about these flies, we may come up with ways to get over these two hurdles, but that's a long way off right now.

In the meantime, if you're heading to the Antarctic, don't forget the bug spray, particularly if you're there in January.

Tonguefish at the Vents

Take one fish. Place in hot water (as close to boiling as possible). Add lots of salt. No, this isn't the latest dish from Jamie Oliver; this is the recipe for what may be the most extreme environment where a vertebrate animal has ever been found. The environment? The mouths of underwater volcanoes in the deepest part of the Pacific Ocean, known as the Mariana Arc. The animal? A type of creature called a tonguefish (because it looks like a human tongue). The discoverers? A team of marine biologists including Dr. John Dower, from the University of Victoria, British Columbia.

Describing it as hot and salty hardly does justice to the environment where the fish are found. First of all, the volcano tops are about 500 metres (1,600 feet) beneath the surface of the ocean. And they're active volcanoes, meaning they're rumbling away and ejecting molten sulphur and, in some cases, molten lava. For sulphur to be in a liquid state like this, the surrounding temperature has to be somewhere between 200 and 240 degrees Celsius. Because this is happening deep in the ocean, the water temperature drops off quite quickly, but it's still at least 80 degrees Celsius, and maybe even closer to the boiling point of water. These pools of sulphur are something to see, says Dr. Dower. "On the bottom of the sea mounts, you have this pool of molten sulphur that might be as much as five metres [sixteen feet] across. It's roiling and boiling; it looks like boiling asphalt."

This is hardly the kind of place where you'd expect to find anything living. When Dr. Dower and his colleagues first went there, he says, "We were using a remotely operated submersible, and we noticed that the bottom was a mottled purple colour and there were rather odd-looking sediments and sand. So as we zoomed into the top part of this volcano, we got close enough to see that the mottled purple came from the fact that the bottom was just completely covered with these small, flat fish."

These fish are no more than about ten centimetres (four inches) long, and, other than their shape and their ability to withstand extreme heat, what's amazing about them is how many of them are found around these pools. In some places they've been found at a density of 300 per square metre (or square yard), although the average is 100 per square metre. And they're living right up against the pools of sulphur. This was a huge surprise, Dr. Dower says. "We were watching this molten sulphur pool, and the geologists were getting really, really excited because no one had ever seen one before. Then [something was spotted in] the corner of the screen, and someone said, 'Hey, what's that?' We zoomed in with the camera, and he said, 'My God, it's a flat fish!' And it's sitting there on top of this molten sulphur that's roiling and boiling, and it looks happy enough. It moved around a little bit and swam off to the edge again and was replaced by another fish that swam out onto the surface of the sulphur pool."

These fish are not only able to survive in extremely hot pools of sulphur, but they seem to be thriving. It's not just heat that's an issue here. The chemistry is also quite disgusting, at least to us. Carbon dioxide is venting from the volcano into the water column, along with sulphurous plumes. In fact, when a normal fish swims over the top of these vents, it is suffocated, or, as Dr. Dower puts it, "narked," by the gases. This usually happens around dawn to fish that have accidentally drifted over the volcano at night. But the tonguefish are unaffected by the noxious gases.

The "narked" fish end up as chow for the tonguefish. Like everything else with this animal, this was a surprise. Early observations had suggested that they would be living on bacteria and small worms, but watching the other fish fall showed the research team how much the tonguefish relied on this alternative form of food.

The tonguefish leaves researchers with almost as many questions as answers. The water is obviously toxic: it's killing any other species of fish that strays into the water column. And the temperature is well above the maximum seen for other species. Even in places where fish live close to hot water, they never seem to get as close to the source of the heat as these creatures do.

It shows, once again, that whenever we think we have nature figured out, something new and weirder than ever comes along.

8

THE GAMES RESEARCHERS PLAY

Scientists are a curious bunch, and occasionally, the only way they can answer the questions they've set themselves is to get right in there and play. It can make their subjects, and the researchers, behave in odd ways...

SPITTING COBRAS

The Wild West was full of legendary gunslingers: Billy the Kid, Annie Oakley, and Wild Bill Hickock, to name just a few. But none of them would have had a chance against the finest sharpshooter of them all. Fastest on the draw, with an accuracy that would make any of them jealous, is the spitting cobra (*Naja pallida*). One researcher who's faced down this sharp-shooting serpent and lived to tell the tale is Dr. Bruce Young, a neuroanatomist at the University of Massachusetts, Lowell.

The spitting cobra looks just like the kind of cobra you see in a movie, the kind with the hood that sways from side to side as a snake-charmer plays a flute. But you wouldn't want to be the person with the flute if the cobra were a spitting cobra. When these snakes get agitated, they spray venom. And this is no wimpy little squirt, like you'd get from the flower on a clown's jacket. These beasts spray more than two metres (six feet)! From the snake's point of view, it's as good as injecting its victim. All venomous snakes have the equivalent of a syringe behind their teeth, but the spitting cobra's teeth are shaped so that the venom goes forward, rather than down. They tilt their heads back, and out squirts the stream of venom.

Dr. Young was the first to notice something odd about the cobras in his lab. "The snakes I had were fairly testy, as a rule," he says. "So whenever I walked in my snake room, they would start spitting venom at me." Before you get too concerned, there was a Plexiglas window between Dr. Young and the snakes. And it led to an interesting observation. "I noticed," says Dr. Young, "that when the spit hits the glass, it makes a very distinct sound. It sounds like a squirt gun hitting a plate-glass window. And when the venom hit the glass, it would leave some really beautiful geometric patterns."

These geometric patterns looked different each time. Sometimes the cobra had sprayed up and down, sometimes side to side, and sometimes in a figure eight. This led Dr. Young to wonder why the snake made such patterns. He travelled to South Africa, where the snakes are found in the wild, to find out. Here's what he did, in his own words. "I would taunt these snakes a bit, and right as they were getting ready to spit at me, I would hold a little Plexiglas shield up in front of my face, and it would catch the spit. And, sure enough, if you looked at the Plexiglas shield, you would see the same kind of geometric patterns that I'd seen in my lab."

Yes, that's right, he taunted snakes. A brave man – with only a small piece of Plexiglas to protect his eyes. Getting venom on your skin is not so bad; unless you have an open sore, it's not going to cause any serious damage. Although, as Dr. Young says, "I've had a lot of venom shot up my nose and in my mouth. It's not a pleasant sensation." Get it in your eyes, though, that's another story. "It is extremely painful in the eyes," says Dr. Young. "Just instantly, debilitatingly painful. And if you can't get it out quickly, it will ulcerate the corneas and can lead to permanent blindness." So, you want to protect your eyes.

Just how does Dr. Young taunt the snakes? He is definitely not playing flute music. He says, "Cobras, as a rule, don't like the appearance of other cobras nearby and displaying near them. So one of the things I'll do is I just make a little cobra shape out of my hand. I bend my hand over at ninety degrees, so my hand becomes like the hood of a cobra, and I'll just wave this at the cobra. And this is often enough to get him to stand up and pay attention. And then, what I do is I charge the snake. I move at it quite aggressively, low, and then stand up quickly, so I'm towering over it." This is definitely a scientist who's willing to take the occasional risk.

Back to the actual experiment. Just like the cobras in the lab, the wild snakes were creating these beautiful geometric patterns on the Plexiglas shield. Dr. Young reasoned that there was either something special about the fangs of these serpents, or they were moving their heads to generate these shapes. When he took pictures of the cobras mid-spit, using a high-speed camera, sure enough – they were wiggling their heads around. Not only that, but the pattern of the venom on the glass matched the way the head moved. When the snake shook its head up and down, there were vertical streaks on the glass. When its head moved in a figure eight, then that was the venom pattern he'd see. That left Dr. Young with the question: why?

Finding the answer required more taunting of the snakes, and this time, Dr. Young wore a special apparatus on his head. He added accelerometers to the side of his safety glasses (at least he wore safety glasses) to track the movements his head made while he taunted the snakes. What he found was surprising (and slightly unnerving). "What would happen is," he says, "if I move to the left, the snake rotates its head to the left. If I move up, the snake rotates its head up. He just follows me. And when I make a movement that convinces him to spit, he accelerates and he wiggles a little more, and he actually patterns the venom where he thinks I'm going to be."

An analogy is a quarterback throwing a football. He doesn't throw it where the receiver is; he throws to where he expects the receiver to be at the end of the play. The snake is anticipating where the eyes are going to be at the time the venom reaches them. Remember, the snake is trying to get the venom into the eyes, to disable its victim.

The snake's stats? They're on the money between 85 and 90 per cent of the time. Which beats most NFL players.

Worm Grunting

You can attract the attention of a lot of animals by calling them. The Pied Piper of Hamelin lured the city's rats (and, in the end, its children) out of town by playing a flute. Pigs respond to a hog call and ducks to a duck whistle. And with cats, you can make any sound you want – it won't matter. They're not likely to come when you call unless it suits them. But the animal that has attracted scientific interest for its response to sound is the worm. Anecdotes have existed for centuries about the art of worm calling, or worm grunting, as it's known, as a way to get worms for bait. Dr. Jayne Yack, a biologist who specializes in animal acoustics at Carleton

University in Ottawa, wanted to know if there is any science behind this art.

How do you actually grunt for a worm? It's not as if the worms themselves make a noise that humans can imitate. To find out, Dr. Yack went to Florida where, each year, there's a worm grunting festival in the town of Sopchoppy. There, she learned the technique for grunting worms. At first, she was actually skeptical that it could be done, but, she says, "I couldn't believe my eyes when we finally went into the forest and saw this happening. [Worms] literally jump out of out the ground when you grunt."

Worm grunters use two pieces of equipment. The first is a stob, which is a technical term for a wooden stake that's about forty centimetres (sixteen inches) long. This they drive into the ground. The second device is what's called a rooper. Dr. Yack describes it as a long metal object. "It could be a saw or a leaf spring from an automobile." Now it's time to grunt. Gently draw the rooper over the top of the stob, like drawing a bow over a fiddle. (This technique is sometimes called fiddling rather than grunting.) The stob starts to vibrate, and it makes a deep whooping sound that rises slightly in tone during the "oo" sound. And, Dr. Yack says, the worms pop out of the ground.

Why does this work? The working hypothesis was that it has to do with the vibration the stob makes in the soil when it's rubbed. At the festival, Dr. Yack discovered, "Some worm grunters are very, very particular about how fast you stroke the stob and rooper, the size and the type of tree that you use [for the wood]." All of these variables affect the strob's vibrations, and may influence the reaction of worms, if vibrations are what they're responding to.

Dr. Yack and her students took their measuring equipment down to Florida and examined whether there was any relationship between the vibrations and the worms surfacing. Sure enough, close to a vibrating stob, worms pop to the surface like mad, but the farther from the stob, the fewer the worms.

Interestingly, Dr. Yack is not the only scientist looking at this question. Who would have guessed that after decades of no interest, all at once, two researchers would study grunting? Dr. Ken Catania, a mole researcher from Vanderbilt University, Nashville, has come to the same conclusion: that worm grunting works because the animals are responding to the vibrations. Being a mole man, Dr. Catania subscribes to the theory that the reason worms surface when the ground shakes is because they're avoiding moles. When moles dig through the ground it causes the soil to vibrate, and the worms, not wanting to get munched as lunch, make their way up to the surface. There's a good reason why: moles can eat their body weight in worms every day. Sure enough, when Dr. Catania played the sound of moles digging to worms, up to the surface they came.

There is another hypothesis, though – that worms are trying to avoid drowning when it rains. The suggestion is that the vibration of the raindrops on the surface warns the worms, and up they come. However, the evidence for this theory is shaky, so we'll stick with the moles.

There's a wrinkle to the mole hypothesis, though. Worm grunting works only on one type of worm, *Diplocardia mississippiensis*, a worm native to Florida. Dr. Yack has tried to repeat this work in Ontario, where there are both moles and worms, but with no success. She's not giving up, though. There are about seven thousand different worm species in the world, and she's determined to figure out which ones respond to grunts and which don't care. By figuring that out, she hopes to learn more about how worms respond to vibrations, which, as a blind animal, must be important to them.

So, if you're out for a walk in the country and you see someone stroking a saw across a stake planted in the field, don't assume she's crazy. She's probably a scientist trying to coax a worm out of the soil.

Flatulent Fish

Every action generates a reaction, and the awarding of Nobel prizes, for significant accomplishments in six fields, half of them scientific, is no exception. The Nobels have inspired the Ig Nobel prizes. The Ig Nobels recognize work that, in the organizers' eyes, "cannot or should not be reproduced." And just as a few Canadians have been awarded the Nobel for their work, we've won some Ig Nobels, too. One was given to Dr. Ben Wilson, currently a lecturer with the Scottish Association for Marine Science, for his work a few years ago at the University of British Columbia, called, "Pacific and Atlantic Herring Produce Burst Pulse Sounds."

Even if you've fished for herring (*Clupea harengus* and *Clupea pallasii*), there's a good chance you've never heard "burst pulse sounds." They sound a little as if the herring is creaking, almost bubbling away. It's an odd sound, to say the least, and one that even Dr. Wilson wasn't expecting. He and his co-authors "were doing a completely different set of experiments, and we were playing sounds to herring and seeing how they responded. And just to monitor whether my equipment was working, we had a microphone in the tank. Not only were the sounds coming from my speaker, but this sound was coming back, and that's what started this whole escapade."

Dr. Wilson wanted to figure out where the creaking noise was coming from, but immediately faced a challenge. The herring were producing the sound only at night, which meant he couldn't see what was going on. He tried a number of different experiments in order to rule out different options. He'd read that some fish could produce sounds with their swim bladders, an air-filled sac inside most fish, so he figured out how to test that. "I took a dead fish," he says, "and gave it a good squeeze, just to try and push that air around and see what it is. And, hey presto, it produced exactly, or almost exactly, the same sound. We ended up filming these fish with low-light-level cameras, and we saw that they produce these sounds by expelling bubbles from what's called the anal pore, which is an exit right next to the anus on the fish."

Yup, "burst pulse sounds" is just a polite way of saying that fish fart. Now, to be fair, it's not a true fart, since the air is coming from the swim bladder, not from the digestive tract, but the effect is the same, if less odorous. Dr. Wilson does have another name for the sound that's specific to these fish. He says, "We came up with a more specific name, which was 'FAst Repetitive Tick' sounds. And the acronym for that is FART."

How much control the fish have over these FARTs is unclear, and why do they make this noise in the first place? Dr. Wilson has tried various methods for scaring the herring but hasn't seen any change in the sound patterns. The only variable he's been able to link to sound levels is the number of fish in the group. The more fish, the bigger the number of FARTs recorded. This suggests that the sound may be a form of communication between individuals, but that has yet to be proven.

It does make sense when you look at the herring's lifestyle, though. As Dr. Wilson says, "They're a highly social fish. They live in enormous schools at sea. They're extremely abundant, and they've been very important for us in terms of fisheries. And during the night, they rise up in the water to just under the surface, and those schools break up, and form these segregated groups." How the groups stay in touch has been a long-standing question, and it's possible the FARTs are the answer.

How do they get enough air to make the sounds? The best guess is that they fill their swim bladders by gulping air at the surface. Then, they'll have enough air in their swim bladders to keep their position in the water and still send out their messages.

So, if ever you find yourself out on the ocean, late at night, on the deck of a boat that's in the middle of a giant school of herring, turn off the engine, sit back, and listen. Who knows, maybe you'll hear the sound of thousands of fish, all letting it rip at the same time. That would truly be a special moment for science.

FEARLESS IGUANAS

"We have nothing to fear," said Franklin D. Roosevelt, "but fear itself." It may be sage advice for us in times of war or worldwide economic depression, but it's not as useful when you're talking about nature. In a world full of predators, it may be that we have nothing to fear but losing our ability to feel it. Fear is a useful adaptation in a dangerous world, as without it, you may not recognize threats to your own survival. According to Dr. Michael Romero, a biologist from Tufts University in Massachusetts, this is the problem marine iguanas (*Amblyrhynchus cristatus*) face in the Galapagos Islands. They've lived for millennia in a land free from predators, and, as a result, they've lost their sense of fear. This may lead to their demise now that the world has come calling on these islands.

These iguanas, famously first described by Charles Darwin, are tame. Iguanas on one of the more remote islands, where they may never have seen a human, have no fear of us. You could walk right up and pick up an iguana. Dr. Romero says, "Even ones who have seen us, you can get within two to three metres [two to three yards] of them without them running away."

This is a different behaviour from iguanas found on the mainland. There, they will run away at the slightest provocation. They certainly won't let you get in close enough to pick them up.

For Dr. Romero, the question was what was going on at the hormonal level in these animals. A number of different hormones, including adrenaline, glucocorticoids, and corticosterone, are produced when animals are under stress. These hormones are important – they help animals (including us) to escape from predators, and may even help protect against injury and disease. What Dr. Romero wanted to know was whether the stress hormones were produced by marine iguanas when he harassed them.

And harass them he did. First of all, he'd approach them. Since these iguanas had seen him before, when he got too close for their comfort, they would run away. But then, he says, "We would chase them around for about fifteen minutes, and just never quite let them settle down, and then we would catch them. Then we would [take a blood sample to] see whether or not they had a hormonal response to being chased around." They didn't. Unlike almost any other animal in this situation, the iguanas produced no changes to their hormones.

Not that they didn't have the hormones. Dr. Romero's next step created even more stress for the giant lizards. "We caught these animals in order to take their blood samples, and then we stuck them inside an opaque bag," says Dr. Romero, "this little cloth bag, for about half an hour. And they didn't like that, as you can imagine. I'm sure these animals thought they'd be about to be eaten." Now he did get a stress response in the hormones.

You might think these marine iguanas would avoid people from that point onward. But no. They did get a little more reactive.

They'd run away if the researchers got within three metres (ten feet), as opposed to before, at two metres (six and a half feet). This is not exactly a dramatic change. The only good news is that now that a stress response in the hormones has been proved, there's a chance that these animals can learn to worry about predators.

It does leave Dr. Romero concerned, though. As he puts it, "Even three metres is probably not enough to survive a bounding cat or dog." For the marine iguanas, this is a serious concern. On one island, the population is being eaten by dogs that were brought by settlers 150 years ago. And they're not the only animals in this circumstance. History is full of examples of island species that have been fine for generations, then the introduction of one cat or dog has led to the extinction of an entire population. Until this study, no one's really known why that was the case. Now we know that while animals don't necessarily lose their physiological stress response, over time they can lose the sense of when to trigger it.

So don't be afraid of fear – embrace it!

Running on Water

You've probably heard a version of this one before: Why did the basilisk lizard (*Basiliscus plumifrons*) cross the river? To get to the other side, of course! It's an old joke, but what's new and amazing is how this little lizard does it. It doesn't swim, crawl along the bottom, or find a bridge. No, this Central American creature walks right across the surface of the water. That's not remarkable for an insect to do, as plenty of them can support themselves on the water's surface. For a creature as heavy as a reptile, however, it's astonishing. While this feat has been witnessed many times over the years, it took the work of Dr. Tonia

Hsieh, a biomechanicist at the University of Florida, to figure out exactly how the animals were doing it.

The plumed basilisk, also known as the green basilisk, is a pretty green lizard, as the name suggests. Young ones are four to five centimetres (one and a half to two inches) long; the adults can get up to forty centimetres (almost sixteen inches) long, not including the tail, which can be as long as the body. And all of them, old or young, can walk on water. "It's quite spectacular," Dr. Hsieh says, to watch the young lizards walking on water. "Little hatchlings will actually bounce across the surface of the water. A lot of times, both their feet are out of the water, not touching anything, so they have an aerial phase. The large ones are a lot more messy. They don't do quite as pretty a job, but they manage to get across as well."

The lizards are moving fast, so you can't really see the feet themselves. What you do see is their arms (the lizards are standing on their hind legs as they run), which windmill frantically, as if to help the animals keep their balance. Film this and slow it down, and you can see their hind feet sweeping backwards across the surface of the water. It looks, frankly, comical.

There's a cartoon aspect to it, too. If they stop, they sink. They also have a tendency to trip, which results in a face plant. But the lizards soldier on, swimming away as if nothing ever happened. And, since they're very good swimmers, and can stay underwater for half an hour at a time, falling in is no big deal for these creatures. This running on the surface is technically known as a "dynamically stable" situation. It's like riding a bike. As long as you're moving, the bike stays up. Stop, and if you don't put a foot down, you're going to fall over.

If they're swimming and want to get back on top of the water to run, they can do that, too. Dr. Hsieh says, "They take their feet, they move them up as high as they can, and they pretty much clap them together underneath them. It's that sort of motion. They pop themselves right back out of the water and just run off."

Why run rather than just swim? Dr. Hsieh thinks it's to get away from predators. Or, in some circumstances, it's just a faster way to get from one place to another. In her lab, she has a water track set up, and she says, when she left the basilisks in with the track, "I would frequently put them on one side and turn around, and they'd be running across and checking things out and just sort of wandering around in the tank, running back and forth across the water."

What are the mechanics behind this feat? The key point for the lizard is staying upright. That might sound easy, but there's no ground under the lizard and the water is constantly moving away fast, so what the creature is really doing is starting to fall and recovering. Which means the foot movements have to be fast, and they have to hit the water hard.

There are two parts to the foot movement. First there's a slap. That's when the foot of the lizard goes straight down, with only a slight movement away from the body, and hits the water. This slap generates a lot of force downward against the water surface, as much as the lizard's own body weight, which means, thanks to the rule about equal and opposite reactions, the body is pushed up. Then, after the slap, there's the sweep, when the foot goes backwards behind the body. Again, strong forces that counter the falling and keep the lizard upright. To picture this, just think of the front crawl, done with your feet rather than your arms.

This wouldn't work for you, though. The lizards have one big advantage over us, beyond just being small. They have huge feet in comparison to the length of their legs. "If we wanted to run across water," Dr. Hsieh says, "our feet would probably have to be about the size of a football field, which isn't very realistic or likely."

We also have legs designed for the wrong kind of physics. On land, running involves a hard surface, and the net result is that each time we hit the ground with our foot, our legs stretch a little, and the muscles store energy, just like a stretched rubber band stores energy. We then use that to push off and take the next step. In the water, there's no resistance, so the muscles don't store any energy. We'd get tired quickly if we tried to use our legs this way.

Too bad, really. Imagine all the places we could walk if we had the basilisk's skills. It would have made medieval moats a waste of time, though.

WASP FACES: FRIEND OR FOE

You've probably had the experience of running into someone you've met before, but you can't remember his name. You aren't even clear where it was you met him. But one thing is sure: you know immediately whether he is a friend or foe. It turns out we have an incredible memory for faces, better even than our memory for names or events. Which makes sense: being able to recognize our friends and enemies is a pretty vital skill. As remarkable as this ability is, though, we're not the only species that can read faces this way. Dr. Elizabeth Tibbetts, an evolutionary biologist from the University of Michigan, believes that paper wasps (*Polistes dominulus*) share our facility for discrimination.

Can wasps even recognize different faces? Luckily, they can. And the paper wasp is a great species to study this in. Each wasp has a unique pattern of black and yellow stripes on its face. In an elegant experiment, Dr. Tibbetts painted the faces of wasps, sometimes blacking out the yellow stripes, sometimes adding more, to see how the wasps around them responded. Sure enough, as soon as its face changed, the wasp was treated like a stranger, and it had to re-establish its place in the wasp community.

This facial recognition alone is quite interesting. Most research on insect senses involves their response to chemicals, since we know insects rely heavily on chemical cues to communicate. The ability to pick out individuals visually, based on the unique pattern of facial markings, is something that most of us would think was limited to creatures with much bigger brains than insects. Nonetheless, these wasps are extremely adept at recognizing one another.

Armed with this knowledge, Dr. Tibbetts moved on to a second question. How long does this recognition last? Other invertebrates, crabs and ants, have been shown to recognize individuals when they meet them, but the memory is very short-lived, no more than a day in most species. Dr. Tibbetts says, "We thought wasps were smarter than that."

And it turns out, they are. Dr. Tibbetts and her team collected female wasps from widely separated locations to make sure they had never met before. Then, they put two of them together. These females, not surprisingly, would start to fight. It was not surprising because these female wasps were at the stage of their lives when they are what Dr. Tibbetts calls "nest-founding females." She says, "In the wild, when wasps emerge from hibernation, they fly around and fight with each other to establish dominance. And sometimes they end up starting nests with wasps that they've fought with before." That's what was going on here. The females were fighting to figure out which was the wasp most likely to lead if they formed a colony together.

Except they never got the chance. Once the rumble was over, Dr. Tibbetts separated the wasps and put them in back in their original cages, among wasps they already knew. Then, after a week, it was time to go back, head to head, with the female from the earlier encounter. This time when they met, "They were much more friendly," says Dr. Tibbetts. "They tended to hang out next to each other and groom instead of biting each other and trying to grapple with each other, or sting each other. Things are much, much calmer on their second meeting." This suggests to Dr. Tibbetts that they remember each other. After all, fighting is dangerous, and most animals avoid it if they can. If these two wasps were remembering each other, then they already knew which was the more dominant, so it wasn't worth fighting again.

This isn't just a case of their growing gentler as they age. Show these same females a stranger, and it's all business again. They can remember a face for up to a week, which is a long time for a wasp.

At this point, we don't know the limit of how long wasps can remember a face. The best hint we have, actually, is how they interact with their relatives. When the females begin building nests in their second year, they'll sometimes collaborate with their sisters and sometimes team up with wasps they've never met before. Dr. Tibbetts says the way they interact changes depending on whether it's siblings or strangers they're working with. While we don't know how they're recognizing kin at this point, it's possible they're remembering facial markings.

What are they using the skill for? Probably to avoid fights. For a wasp, every fight comes with the risk of death, so it makes sense that she would stay out of squabbles whenever possible. Just like humans. Or not.

Fruit Fly Fight Club

Most of us associate the fruit fly (*Drosophila melanogaster*) either with genetics or rotting bananas – or both – but not with fall-down, body-slamming, no-holds-barred fighting. And yet, Dr. Ed Kravitz, a neurobiologist at Harvard Medical School, has created a fight club for fruit flies in his research lab. Not only that, he's discovered that male and female fruit flies have different rules for fighting.

In the hundred years of fruit fly research, there have only been a handful of papers on the subject of their fighting, and most scientists aren't familiar with the behaviour. That's because in the lab, it's not something they see very often. Fruit flies don't fight unless they are really hungry. Limit the food and the gloves come off, so to speak.

When two male fruit flies face off, first they'll spread their wings, to try to intimidate their opponent. It's the fly equivalent of trash talking. Then, the action begins. If the two males are really going at it, as Dr. Kravitz puts it, "They get up on their hind legs and they duke it out." They also perform a manoeuvre he describes as lunging, where they rise up and then body-slam down onto their opponent. He says, "It doesn't look like much when you see it in real time, but when you look at it in slow motion, you find that they actually flatten the opponent when they hit them."

Females take a different approach. Dr. Kravitz calls their moves "shove and head-butt." Put a female close to some food (yeast paste is the favourite), and if another female comes too close, the first will try to push the intruder away with her front legs. If that doesn't work, she head-butts it. Yes, she bangs away at her opponent with her head.

Dr. Kravitz didn't stop at discovering this difference between females and males. He's now figured out how to change the rules and make males fight like females, and vice versa, by manipulating their genes. There's a gene in fruit flies called the "fruitless" gene, which is essential for courtship and mating. In early experiments, researchers found two forms of the gene, one for males and one for females. Then they started experimenting, by putting the female form into male flies, and vice versa, with the result that the males tried to court other males, and females other females. This discovery made the papers a few years ago, and in some circles it was hailed as a "gay gene" (although it was later shown to not be the whole story).

Where does Dr. Kravitz's work fit into this? He took the same manipulated flies and looked at their aggressive behaviour. Sure enough, the males fought like females and the females like males. Which tells us that aggression, in fruit flies at least, is controlled by the same genes as gender.

The ultimate question, of course, is whether this applies to us. The short answer is that we have no idea. Dr. Kravitz argues that there is likely a genetic influence on aggressive behaviour in all animals, but that, in humans at least, the way we are raised has a huge and incalculable effect on how aggressive we can be.

The next time you have fruit flies in your kitchen, though, don't hide all the food on them. You might just start a war.

TADPOLE BAIL OUT

The egg is often said to be nature's most perfect design. It's hard to break and gives the embryo everything it needs to survive and grow. But there are some limitations. The youngster inside can't move the egg, which is a problem if a predator comes along. The red-eyed tree frog (*Agalychnis callidryas*) has come up with a unique safety system. The young tadpoles can bail out when they sense a predator approaching. Dr. Karen Warkentin, a biologist at Boston University, has figured out how the tadpoles perform this particular feat.

In recent years, the red-eyed tree frog has become the poster child of the rain forest. It is a bright green frog, 5 to 7.5 centimetres long (2 to 3 inches), with red eyes, often with orange feet and blue and yellow stripes on the side. It has an interesting life cycle. The female lays her eggs in gelatinous clutches on the undersides of leaves, rather than directly in the water, which is the norm for most frogs. When the embryos form into tadpoles, usually after about a week, they wriggle free from the eggs and drop into a pool of water below. They spend two to three months in the pool developing into frogs, and then climb up into the trees, where they live as adults.

At least, that's the normal pattern. But Dr. Warkentin has established that these animals will sometimes drop from the eggs into the water two or three days early, if they find themselves under attack from a predator. In their native Costa Rica, that might be an egg-eating snake, or certain species of wasp. Since the tadpole is not fully developed, this is something of a risk for them, but it's better than the certain death waiting for them if they don't jump. They're still easier than a mature tadpole for predators to pick off in the water, but there's a chance they'll stay alive. They can react to threats very quickly. Just a few seconds after a snake attacks the clutch, tadpoles will start dropping into the water. It looks quite spectacular, says Dr. Warkentin. "You get a snake with its face stuck in the clutch and embryos falling down around it. Sometimes, you get eggs that are hatching even in the snake's mouth and then falling down, or it could be that the snake takes a bite and rips it off and then the rest of the clutch starts hatching out, and the snake is kind of looking at it. I think they look kind of surprised sometimes."

Dr. Warkentin wanted to know how these embryos, which have no experience of a snake attack, know that bailing is their best choice. From the start, she thought it had something to do with the physical disturbance of the eggs, because, she says,

"They don't hatch if a snake is looking at the clutch. It's actually only when the snake starts biting on an egg mass that the embryos start to bail out. So it looked like it was something about that physical disturbance to the clutches, as opposed to a visual cue or a chemical cue."

To test this hypothesis, Dr. Warkentin started recording vibrations in clutches under attack by snakes and also the vibrations from rainstorms. Then she played both types back to the eggs in her lab. She says that she essentially asked them, "'Does this scare you? Does it induce hatching?' And I found that the snake vibrations did induce hatching. So without an actual snake anywhere in the vicinity, just playing them the feeling of the way that they would be shaken by a snake attack induced the embryos to hatch."

It's important to note that vibrations from rainfall did not have the same effect. This means that the eggs are tuned to the snake vibrations, which are quite different from the tremors caused by rain. A raindrop hitting a leaf is a quick event, while the snake, as Dr. Warkentin describes it, "is kind of chewing and scuffling around in there – even a short snake can insert its face and pull back the eggs – whereas a raindrop is more like a simple tap." There are also different frequencies for the two kinds of disturbances, with the snake's being much lower.

Being able to identify the snake's rumble is very effective. Usually about three quarters of the clutch will survive by bailing out before they're eaten. How many last in the pool until maturity isn't known, but likely some do. They are one smart kind of embryo. We tend to think of embryos as inert, waiting for hatching to get going. But in this species, its first life-or-death decision sometimes happens before birth. Now, if only someone could get a microphone in close enough to listen and hear if any tadpoles call out "Geronimo!" as they fall.

Flying Ants

You've probably heard of flying fish. Flying squirrels, too. Even flying foxes. But have you ever heard of skydiving ants? In 2005, that was a new one, even to biologists. Dr. Michael Kaspari, a researcher at the University of Oklahoma, and his colleagues, were studying canopy-dwelling rain forest ants (*Cephalotes atratus*) when they encountered these tiny diving daredevils. These ants live in the treetops of South and Central America, in the sunny, leafy ecosystems that exist thirty metres (a hundred feet) above the forest floor. At least they do when they're not taking plunges that would qualify them for the next Insect Extreme Games.

Even the look of these ants is pretty extreme. They're about a centimetre (less than half an inch) long, but, as Dr. Kaspari describes them, "They look like dinosaurs, all long legs and spines all over their body." The spines are important protection for the ants against getting eaten by birds and reptiles. But, if you were to grab one of these ants, the spines would make it stick to you.

The discovery of their skydiving was an accident. One of Dr. Kaspari's colleagues, Dr. Steven Yanoviak (we met him earlier, in the story of the parasites that make some ants look like red berries), was up in the canopy one day when he brushed some ants off the tree. As he watched, the ants somehow managed to alter their course midair and land on the trunk. The next thing you know, a research paper was born.

Of course, you can't just look at ants falling in the forest and call it a study. No, the team started looking at them in more detail. It started simply enough. First, take a small weight that's about the same mass as an ant, and drop it from the canopy. Watch it fall, and see what happens. Compare that to the ant. In the absence of any breeze, the weight, of course, drops straight down, but the ant would drop down about five metres (sixteen feet), then make a mid-course correction and head towards the tree.

Why wait the five-metre drop? It's because the ants can't start to control their fall until they reach a minimum velocity. But at that point, they can start to twist and move and choose how they fall. That's when they head back to the tree trunk. "It's amazing," says Dr. Kaspari. "When these experiments were done, oftentimes you'd see the same ant crawling right back next to you on the branch where you dropped it, so they're pretty good at coming back home."

And they do this without wings. Their parents, the queen and the males, had wings during their mating flights, but all the worker ants, who are the ones doing the gliding, do not have wings and never have had them. But, Dr. Kaspari says, "Sometime in the past, the ants started colonizing the canopy, and suddenly they encountered something that they never had encountered in evolutionary history before, and that is a long fall. And so, amazingly enough, they took all that neural circuitry from their parents and managed to find a way to glide back to the tree trunk." Those that didn't make it would have been quickly selected out of the population.

Biomechanicists, the researchers who study this kind of behaviour, have a scale of animal flying. Dr. Kaspari describes the range: from parachuting, which, he says, "is essentially holding your legs out and praying as you fall, to a kind of a directed descent in which they actually show a fair bit of lateral movement away from a straight plummet to the earth, to gliding, which is the most lateral, controlled form."

These ants make a directed descent. It's true gliding, as they only go down, but at least they have some level of control and can change the angle of their descent to be almost horizontal to the ground. And they do this whole thing backwards. When they hit the tree, it's "gaster first," which, in English, means butt end first. That might sound strange, but these ants are falling at fourteen kilometres (eight and a half miles) per hour, fast enough that it's probably not a good idea to hit their heads when they contact a tree. The ants have claws on their back feet, and they are light, so when they hit the tree, they hang on with their feet. If they miss, they just bounce off and try again. Living in the trees, they need to be able to grip quite well anyway, so it's no surprise they can hold on when they're landing.

The reason for their skydiving seems to be the same as the reason for their spines. The treetops are, in Dr. Kaspari's words, "a jungle, and these ants encounter all sorts of predators, including lizards and birds and perhaps even army ants that come up and look for them. So what may have happened is that an adaptation, which first was good for getting back if they accidentally fell, became an ace-in-the-hole to escape predators. So that, if a lizard comes after them, the last thing the lizard will see is the ant looking up at him as she sails back to the tree trunk, maybe five or ten metres [fifteen or thirty feet] below."

Better than climbing all the way up the tree from the ground.

David Elliot

Pat Senson is also known as the Science Guy on CBC Radio One's local drive-home shows across the country. A trained biologist, he was for many years a producer with the multi-award-winning science program *Quirks & Quarks*, the oldest radio science program in the world. He was co-winner of the American Institute of Physics Award for Best Science Writing in 2003 and again in 2007. As a science journalist, he has covered everything from what existed before the Big Bang, to what will happen at the end of time itself, but he has a special love for living organisms, both large and small. This love was fostered during his days as an undergraduate at the University of Guelph, where, along with his roommates, he decided to create a whole living ecosystem. One wading pool, a handful of terraria, some crayfish, spiders, crickets, and snakes later, a biological teaching tool was born. A few insects escaped, but nothing was harmed in the creation, and it did teach Pat a lot about some of the creepier of the creepy crawlies. Today there are fewer wild animals in his home and more textbooks, as he is now studying law at the University of Toronto. He lives in Toronto with his partner and two slightly addled cats.